Biomass to Biofuels

Biomass to Biofuels

Edited by

S. Syngellakis
Wessex Institute of Technology, UK

WIT*PRESS* **Southampton, Boston**

Editor:

S. Syngellakis
Wessex Institute of Technology, UK

Published by

WIT Press
Ashurst Lodge, Ashurst, Southampton, SO40 7AA, UK
Tel: 44 (0) 238 029 3223; Fax: 44 (0) 238 029 2853
E-Mail: witpress@witpress.com
http://www.witpress.com

For USA, Canada and Mexico

Computational Mechanics International
25 Bridge Street, Billerica, MA 01821, USA
Tel: 978 667 5841; Fax: 978 667 7582
E-Mail: infousa@witpress.com
http://www.witpress.com

British Library Cataloguing-in-Publication Data
A Catalogue record for this book is available
from the British Library

ISBN: 978-1-78466-034-5
eISBN: 978-1-78466-036-9

Library of Congress Catalog Card Number: 2014952566

Printed by Lightning Source, UK.

Preface

Biomass is a continuously renewed source of energy formed from or by a wide variety of living organisms. Through biochemical and thermochemical processes, it is converted into gaseous, liquid or solid biofuels, which already meet a significant share of the current world energy needs. Because of their contribution to the sustainability of energy supply, reduction of green house gas emissions as well as local employment and energy self-reliance, research interest and activity in enhancing biofuel energy output, efficiency and performance remain strong. The purpose of the present volume comprising a selection of distinguished contributions from the Transactions of Wessex Institute is to focus on recent developments in this topical research field.

The first part of the volume comprises five articles mainly concerned with biomass resource potential and management. More specifically, the reported investigations assess grass and lawn substrates, rapeseed straw and microalgae from Upflow Anaerobic Sludge Blanket (UASB) reactor effluents as possible sources of biogas, bioethanol and biodiesel, respectively. The calorific value of biomass pellets produced from a wide range of agricultural waste is assessed in the context of improved environmental and economic management. Research on alternative feedstock substrates proposes optimised cropping systems leading to sustainability of biomass production and environmental benefits.

The emphasis in the subsequent group of eleven articles is on biomass conversion processes, aiming at assessing performance as well as output quality and diversity. Issues addressed include: air pollutants in the form of suspended particulate matter during combustion of waste rice husk and straw; product characterisation and quality control in the case of char production from switchgrass energy crop; catalytic tar decomposition with the concomitant enhanced gas production from the pyrolysis and gasification of cellulose, a major component of wood and plant biomass; design of agro-industrial residue recycling plant for optimum and diverse biogas capacity; conditions for improved efficiency in the liquefaction process of waste bamboo and other woody material; two phase treatment of dairy manure by anaerobic digestion for increased methane production; review of current anaerobic digestion techniques and environmental impact assessment of their application in Europe; new route for non-enzymatic hydrolysis using ionic solutions and comparison with the enzymatic one.

Biodiesel, a fluid biofuel produced from biomass with high lipids such as rapeseed oil, sunflowers and soy beans, is the focus of two articles: the first investigates the effect of biodiesel blending with diesel fuel on diesel engine performance and emissions, the second assesses the efficiency of catalytic reforming of biodiesel into a gaseous mixture, used directly as Solid Oxide Fuel Cell (SOFC) fuel.

In the last three articles, the prospects of biofuels as viable sources of energy are examined within European contexts. The first article addresses and balances the arguments about high costs and low ecological benefits with regard to the promotion of biofuels and future relevance of their use for transport. The second article argues that sustainability dependence on availability of resources, political determination and technology development ensure the quality as well as the further development of the conversion processes. In the last article of the volume, the effect of a biomass plant on environmental balance is assessed through mass and energy balances as well as pollutant dispersion.

In summary, the present volume addresses a significant number of important themes and thus combines subject breadth and density with in-depth study of biomass resourcing and processing as well as the issue of biofuel and renewable energy sustainability.

Stavros Syngellakis
The New Forest, 2015

Acronyms

AD	Anaerobic Digestion
ADL	Acid Detergent Lignin
AES	Atomic Emission Spectroscopy
ATR	Autothermal Reforming
BET	Brunauer-Emmett-Teller (technique)
CCA	Copper, Chromium and Arsenic
CDM	Clean Development Mechanism
CER	Certified Emission Reduction
CGP	Cumulative Gas Production
CHP	Combined Heat and Power
COD	Chemical Oxygen Demand
CSTR	Completely (or Continuously) Stirred Tank Reactor
DCM	Dichloromethane
DH	District Heating
DP	Degree of Polymerization
EC	European Council
EDXS	Energy Dispersive X-ray Spectroscopy
EEA	European Environmental Agency
EIA	Environmental Impact Assessment
EU	European Union
FFA	Free Fatty Acid
FIRSST	Feedstock Impregnation Rapid and Sequential Steam Treatment
FTIR	Fourier Transform Infrared Spectroscopy
GC	Gas Chromatography
GC-FID	Gas-Chromatograph with Flame Ionization Detector
GC-MS	Gas Chromatography - Mass Spectrometry
GC-TCD	Gas-Chromatograph with Thermal Conductivity Detector
GCV	Gross Calorific Value
GHG	Greenhouse Gas
GHSV	Gas Hourly Space Velocity

GIS	Geographic Information System
HMF	Hydroxymethylfurfural
HPLC	High-Performance Liquid Chromatography
HRT	Hydraulic Retention Time
ICP	Inductively Coupled Plasma
JIS	Japanese Industrial Standard
MBT	Mechanical Biological Treatment
MSW	Municipal Solid Waste
NDF	Neutral Detergent Fiber
NHV	Net Heat Value
ODM	Organic Dry Matter
OFMSW	Organic Fraction of Municipal Solid Waste
OLR	Organic Loading Rate
OSR	Oilseed Rape
OUR	Oxygen Uptake Rate
PAH	Poly-Aromatic Hydrocarbon
PEG	Polyethylene Glycol
RDF	Refuse-Derived Fuel
RSM	Response Surface Methodology
RSREG	Response Surface Regression
SCR	Selective Catalytic Reduction
SEM	Scanning Electron Microscopy
SOFC	Solid Oxide Fuel Cell
SPM	Suspended Particulate Matter
SR	Steam Reforming
TDS	Total Dissolved Solids
TG-DTA	Thermogravimetry - Differental Thermal Analysis
TSS	Total Suspended Solids
TVFA	Total Volatile Fatty Acids
UASB	Upflow Anaerobic Sludge Blanket
VDS	Volatile Dissolved Solids
VSS	Volatile Suspended Solids
WCS	Whole Crop Silage
WGS	Water Gas Shift
WWTP	Wastewater Treatment Plant
XRD	X-ray Diffraction

Contents

Energy recovery of grass biomass

S. Oldenburg, L. Westphal & I. Körner
Hamburg University of Technology, Germany

Abstract

Not only due to the actual climate change, but also in consideration of exhaustible resources, alternative energy supplies for the steadily growing energy demand need to be found. The main emphasis should be placed on the substitution of fossil fuels with agricultural by-products and other organic materials. Especially, the utilisation of fresh grass or grass silage of extensively cultivated farm land has great potential as an energy feedstock, as in agriculture this bio-resource is currently considered a waste material and is neither economically nor ecologically utilised in an efficient way. Additionally, the lawn-clippings, which are accumulating as communal and private waste could be used for energy production since many local authorities have problems with utilising this gramineous waste. Thus, for anaerobic digestion big amounts of grass biomass and lawn substrates are available from farmers and landscape conservation. In order to evaluate the suitability of this material for biogas plants in the first place, a detailed inventory needs to be conducted. This was exemplarily done for the Hamburg District Bergedorf (155 km^2). The result shows that approximately 10,000 Mg/a of grass and lawn clippings could theoretically be made available. By laboratory investigations, in batch tests, the theoretical biogas potentials of selected grass and lawn substrates were determined. A statement about the suitability of the substrates for anaerobic digestion is made in this paper. The biogas potentials are between 325 and 720 standardized l per kg organic dry matter (l/kg ODM), depending on the sampling location, mowing time, grass species etc. For example, the biogas potentials for clippings from the dikes were in a range comparable with corn silage between 420 and 700 l/kg ODM. Additionally, the problem of seasonal accumulation of grass biomass including the influence of storage on the initial material is considered in this paper.
Keywords: grass, lawn, digestion, biogas potential.

1 Introduction

Climate protection is an important part of the challenges of environmental politics nowadays. To prevent the progress of global warming, climate protection agreements have been established according to which a great amount of the energy demand has to be provided by renewable energies in the near future. The substitution of fossil energy sources by renewable bio-resources will inevitably lead to less emission of carbon dioxide. Moreover, renewable resources are an important and versatile energy source since they can be utilized for heat, fuel and electricity. A crucial point is the possibility of storage and flexible utilisation. Because of this, the cultivation of energy crops, especially corn, increased in Germany in the last years enormously. But at this point, the competition for cultivable areas between energy crops and food products has to be considered. Eventually other conflicts of interests will rise up between bio-energy providers, nature conservation and tourism, given that much grassland was converted to farmland for growing energy crops. A better opportunity could be the use of plants and green waste that arises from countryside preservations and which cannot be used for the food production. There are an estimated 900,000 Mg of green waste available in Germany per year, which originate from nature and biotope areas [1]. Additional grass fractions are generated in agricultural excess areas or lean grass fields. Usually these resources have to be collected from the land on a regular basis to prevent accumulations of nutrients in the soil and preserve the biological ecosystem. Furthermore, due to the decrease of cattle farming in some regions, the green waste from the farmland formerly used for cattle feed production needs to be disposed of. These areas are then only serviced without utilisation of the grass. The maintenance of public parks, sport fields and gardens are other sources of grass. In Germany an estimated 300,000 Mg/a of such green waste is generated [1]. In conclusion, grass fractions are available in huge amounts from private and commercial sources as well as from landscape maintenance. The goal of this work is an evaluation of grass fractions as substrate for biogas generation. The yield of grass biomass per area has exemplarily been investigated for one district of Hamburg. This energetically usable biomass derives from extensive grasslands, fields of grass and farmland; it is either disposed waste or utilised up to now. The amount of this biomass has been investigated and experiments were performed to determine its energetic potential. The results can be transferred to different areas and regions to draw a final conclusion: whether it makes sense to energetically utilize the unused or disposed grass and lawn waste as a substrate in biogas plants, or not.

2 Definitions

Grassland and pasture is defined as agricultural land that has been set up by humans. The grass has to be cut on a regular basis to prevent a scrub encroachment [2]. Grass, herbs and leguminous plants grow on grassland. Depending on the farming of the land grass usually represents the majority of the present vegetation in Germany. This is due to the selection caused by mowing the land repeatedly, since grass is able to regenerate and re-grow very fast. Thus the ratio of grass increases with intense farming of the land. 29% of the agricultural areas of Germany are grasslands [3]. A field of lawn is defined to be an area which includes many different kinds of grass, but no herbs or leguminous plants at all. To maintain a good optical view, the lawn is usually kept at a length of two to ten centimetres. Therefore it has to be mowed up to 40 times a year [4]. Fields of lawn are maintained for the purpose of sports, recreation and a positive representation. About 5% of the entire area of Germany is kept as lawn of some sort [5].

3 Inventories

The district of Bergedorf covers an area of 155 km^2, which represents 20% of the entire area of Hamburg, including 13 different quarters. The district itself could be split into two different parts: an urban part and a rural part, which makes up about 75% of the area. Most of the quarters are agrarian and have a small population density. Overall there are about 120,000 residents living in the district of Bergedorf, averaging at 760 people per km^2. There are only three quarters with a population density of more than 1,000 people per km^2. The other ten quarters are sparsely populated with no more than 400 residents per km^2. The grassland is about two times as large as the cultivated land with an area of 53 km^2. More than half of the area is being used agriculturally. The area used for buildings, infrastructure, business and the open space add up to only about 26%.

Almost half of Hamburg's agricultural land is located in the district of Bergedorf whereas the cultivated land in the entire city is only about 10%. Altogether all different structures of a city are present in the district of Bergedorf: ranging from urban regions to less densely populated parts and from agricultural land to nature preserve areas. The calculation of the dry mass yield of grass fractions was performed using the land use of the city, values from literature and average values. Additionally all results presented in this paper refer to the generation within one year. Furthermore the dependencies on location, climate and cultivation have been considered in the calculations.

3.1 Fields of grass

The actual yield potentials of the different grasslands are hard to approximate. There are literature values, but the yield can differ depending on the location and nutrient content of the ground. The grassland was categorized into two different main groups, whereat the areas have mostly been assigned to their cultivation. The first group is the intensively managed grassland, which are intensively used fertile meadows in agriculture whose grass is used to feed the cattle. The second group is the extensively managed grassland. This is used in accordance with the administration. The areas are used for conservation or recreation; some of them are defined as compensation areas. The entire potential of the whole grassland is about 8,100 Mg of dry matter in the district of Bergedorf per year. This is

67.5 kg per resident and year. In the following sections, the calculations are shown and explained.

3.1.1 Intensively managed grassland

To determine the potential of intensively farmed meadows, letter-surveys have been conducted. 81 were sent to farmers in Bergedorf via the department of agriculture. The grass from intensively managed grassland is only available for further energetic utilization if the outgrowth could not be utilized in another way. The reasons why the mowing is not used completely, are, for example, a lack of cattle, a bad quality of the year's harvest or plant populations that cannot be consumed by the cattle. Mowing and gathering this kind of grass could only be considered if an allowance was paid for this work. The average yield from mowing three to six times per year is about 0.8 and 1.2 kg of dry matter per square meter and year [6]. For a yield of 0.85 Mg of dry matter per square meter and year there is a potential of about 1,780 Mg of dry matter per year from intensively managed grassland.

3.1.2 Extensive managed grassland

There are five different kinds of grass areas in the district of Bergedorf, which are extensively managed: dikes, public fields of grass, compensation areas, contracted nature preservation and nature preserve areas. The dikes, the public fields of grass and compensation areas are fertilized with liquid manure or dung in part. Then the yield per year and square meter is less than 0.6 kg for cutting it twice a year [6]. Land that is not being used to produce animal food and must not be fertilized, as contracted nature preservation areas have a yield of biomass production less than 0.35 kg of dry matter per year and square meter [6]. Nature reserve areas will be considered to have a yield of 0.15 kg of dry mass per year and square meter [7].

The potential of dikes was computed using the average of the year 2010 and is about 600 Mg of dry matter per year. Most of the outgrowths are disposed by composting, a process that causes costs. Gathering of the material is expensive and the incline of these fields impedes this process as well. The maintenance of the areas that are under contracted nature preservation is tied to distinct contracts with the farmers of Bergedorf and the agricultural use as pasture or mowed meadow is defined. These areas can be mowed starting at the first of July; the responsibility of this action was transferred to the farmers. The farmers are obliged to remove the cut grass for preventing an accumulation of nutrients on the ground. For the purpose of disposing this material it is commonly used as litter for the animal sheds. This is not the best solution and it does not represent the optimum utilization of the material. It takes years up to decades until the land is depleted and there are different meadows with larger or smaller yields. Land having very small yields due to the weather can usually only be cut once a year. Cattle are sent on some of this land after cutting the grass, then the amount from the second cut decreases, too. By the fact that these areas are farmed by ecological means, indicating that the first time of cutting the grass is done very late and a second cutting does not always take place, a potential of about

1,300 Mg of dry mass of grass per year from areas that are under contracted nature preservation is theoretically available.

Notice has to be taken of the partial swamp land and the areas that are hard to access, properties that complicate collecting the material. Additionally when cutting the grass very late they contain an increased amount of lignin, which can cause problems during the biogas production. The maintenance of the present compensation areas in the district of Bergedorf and areas that are under a development plan is performed using similar conditions in the according contracts compared to the contracted nature preservation, but the use of the material from cutting is similar to the contract nature preservation. The mowing is only used occasionally for meadows with no cattle on it. These areas are partially fertilized using cattle dung. So they are considered as extensive grassland having yields of about 0.6 kg of dry matter per square meter and year by cutting twice a year. For grazed fields there is a computed value of 0.35 kg dry matter per square meter when cutting the grass. The second clipping is usually dried and used as litter in the cattle stables in September. This represents a bad compromise, since the quality of the material is not sufficient to use it as food and therefore it would have to be disposed. The theoretical potential was determined to be about 5,400 Mg of dry matter per year from compensation areas.

For the nature preserve areas, proper calculations of the incoming amount of grass are very difficult, since maintenance is only done on rare occasions. The actions of maintenance are integrated into a maintenance schedule individually adapted to each nature preserve area. It is a matter of fact that a large amount of the material cannot be utilized, because a part of it has to remain in the nature preserve areas to preserve the natural circle. In addition, the areas are quite often difficult to access and agricultural vehicles are not allowed to drive there. By an energetic utilization and an according allowance for the outgrowth the maintenance of the nature reserve areas can be improved and a larger potential is possible. The yield is at about 0.15 Mg of dry matter per square meter and year, because of a late cutting and due to a lack of available nutrients in the ground. Hence the potential of the biomass is estimated to be 280 Mg of dry mass per year from nature preserve area.

There are many small companies, who are assigned to maintain public fields of grass. Those areas are only mowed about once or twice a year, so they belong to the group of less intensively managed grasslands and not to the group of lawn. Because of not fertilizing these areas the yield potential was estimated to be 0.35 kg of dry matter per square meter and year. So far the cut grass remains on the land and is not gathered, except for a small path. Currently, it is desirable to get less mowing. Nonetheless the disposed material is being composted which causes costs. From these results a biomass potential of 1,120 Mg dry matter per year from public fields of grass were detected. By mulching after clipping some nutrients are transported back into the ground and the yield could be increased.

3.2 Fields of lawn

In this work fields of grass and lawns will be discussed separately, since their maintenance and composition of different kinds of grass are different. The fields of lawn of public grasslands, parks, sport fields, private gardens, single and apartment buildings have been considered for the calculations. The potential for public grassland with a total of 10 square km has been calculated on indications of the district office Bergedorf, surface structures and re-growth rates. In addition, the amounts which are partly collected are included in the calculations. A total of 490,000 m² of lawn is cut by companies. It is not known how large the surface is which is cut from the District Office (Bezirksamt). It has made the assumption that at least another 20% of the total of 7 square kilometers of public green areas must be mowed. It was assumed that all areas are covered with grass, which has similar properties as the strain "Berlin Zoo". Since the area size is not precisely known, a much larger potential might be possible. With this calculation the potential of public grassland is estimated to 643 Mg of dry mass per year.

A questionnaire was sent to all churches in Bergedorf. The response rate was very low. Therefore, only a reflection of the state church has been done. The result was about 10 Mg of dry matter per year. Knowing that only one cemetery has been considered for calculating the potential, there will be much more grass waste that could not be included so far.

The sport fields in the district of Bergedorf could provide about 32 Mg of dry matter per year, if this material would be gathered and not remain on the field. The potential yield of private yards and gardens of single, double and multiple apartments depends heavily on the settlement structure, the size of the gardens and their design.

In the urbanized parts of Bergedorf the grass potential for the one-and two-family houses was calculated with an average of 34 kg of organic waste per inhabitant per year. This value comes from the estimation of the Ministry of Urban Development and Environment, which is in Hamburg for about 60,000 Mg of organic waste per year recycled about private composting [8]. For the remaining eleven city quarters, this value could not be applied because of their rural structures. Therefore the garden sizes and structures were determined by sending letters and surveys to the residents and by performing green waste sorting. In addition, the re-growth rates and expected care measures were calculated. It is to be remarked that the calculated potential reflects actual and realistic values, if proper collection procedures are introduced and the residents are well informed concerning the separation of different kinds of waste. The computed potential from private households is about 4,400 Mg of dry matter per year. The whole potential of lawn is about 5,070 Mg of dry matter per year.

3.3 Summary

There is an overall theoretical potential of 15.550 Mg of dry matter per year of intensively and extensive managed grassland and lawn.

Table 1: List of the potential yields of grass in the district of Bergedorf.

Area	Dry matter [Mg/a]
Intensive grassland	1,780
Extensive grassland	8,700
Fields of lawn	5,070
Total	**15,550**

4 Anaerobic digestion of grass fraction

Biogas is a mixture which primarily consists of two components: methane (CH_4) and carbon dioxide (CO_2). It is generated by the degradation of organic matter in the absence of oxygen. The technically usable process of this anaerobic degradation is called digestion, which is enabled by complex interactions of different microorganisms. The degradation proceeds step by step and can be broken down into four different states that are referring to the involved microorganisms: Hydrolysis, Acidogenesis, Acetogenesis and Methanogenesis. Being the prime combustive compound, methane basically determines the properties and the energetic usage of the biogas. It is an energy source with a high yield and has the calorific value 50 kJ/kg in under standardized condition and represents about two thirds of the overall colorific value of natural gas [9].

A detailed literature research has shown that the biogas potential of grass varies between 500 to 700 l/kg ODM and is quite different concerning the structures of the land or the agricultural methods. In consequence, the gas potentials of many different grass samples were determined by theoretical models, elementary analyses, as well as by laboratory experiments of exemplary samples of the district of Bergedorf. The calorific value determined by thermal oxidation serves for comparison in section 5. Choosing the particular grass substrates was the priority to the broad range of different areas and characteristics. All of the samples have directly been collected in the district of Bergedorf and cover all areas considered in the inventory. Altogether 16 different substrates have been investigated; nine substrates of extensive managed grassland, four different lawns and one of intensive managed grassland.

4.1 Determining the theoretical biogas potential

Using elemental analysis all ratios of carbon, nitrogen, sulfur and phosphor of the different substrates can be determined. The results are listed in table 2. The content of carbon of the grass samples was between 45% and 47% of the dry matter. The content of nitrogen varied in a range from 0.72% for outgrowths of nature preserves areas up to 37% in case of public fields of grass. The highest value of nitrogen can be lead back to cut grass that has been left on the ground aggregating nutrients.

The ratio of carbon to nitrogen (C:N ratio) is crucial to the gas yield. If the ratio is too small, the organic substance of the substrate cannot be degraded completely. This is the cases for very low crude fibre contents, as well. The optimum ratio is reported to be between 21:1 and 40:1 [10]. For ratios lower than 16:1 the gas yield is expected to decrease, since there is a lot of redundant nitrogen. The maximum ratio should be 45:1. Many samples were in the optimum range, only the samples of dike clipping c and d, as well as the areas in the nature preserve areas had too low nitrogen contents (table 2). The phosphoric content of the grass samples from this district, except for the ordinary maintained grass of gardens and yards, are lower than the suggested literature values of 0.14% to 0.19% phosphor in the dry matter [11]. This increased phosphorous content can be lead back to the fertilization in spring. The chlorine content of the substrates is in a range of 0.23% up to 1.16% in the dry matter. The literature values are 0.31% for festuca arundinacea and 1.39% for grass of pasture and are not exceeded by the substrates of district Bergedorf. In contrast, the sulphuric content is higher: up to 0.3% sulfur in the dry matter. The literature values are between 0.14% and 0.19%. Sulphur is an essential component for microorganisms that are present in a digester. Not solved resp. dissociated hydrogen sulphide is toxically and can hinder the biogas production if it is present in increased concentrations.

Table 2: Elemental composition of the samples.

Substrate	**C**	**N**	**Ph**	**S**	**C:N**
Unit	% TS	% TS	% TS	% TS	
Public meadow	47.1	2.4	0.08	0.30	23.2:1
Dike clipping (a)	46.6	1.9	0.04	0.18	29.4:1
Dike clipping (c)	46.0	0.7	0.10	0.15	69.1:1
Dike clipping (d)	45.6	0.8	0.13	0.19	69.2:1
Nature preserve area	46.8	0.8	0.13	0.16	74.5:1
Grass of gardens and yards (ordinary)	45.4	1.6	0.21	0.30	31.9:1
Grass of gardens and yards (properly)	34.5	1.7	0.10	0.27	25.0:1
Public fields of grass	44.2	1.9	0.13	0.24	27.7:1

One possibility to theoretically determine the maximum biogas potential is to perform calculations including the elemental composition and to use the Buswell-equation that has been modified by Boyle. Regarding that a complete digestion of the organic substance only is possible up to a maximum of 83% the theoretical gas yield results in an average value of 820 l/kg ODM (table 3).

4.2 Practical biogas potential determination

The biogas potential has been determined by batch tests under laboratory conditions according to VDI-standard no. 4630 in triplicate. Digestion test in the batch procedure allow drawing conclusions on the actual biogas yield and the anaerobic degradability of certain matters or solutions. Furthermore a qualitative estimation of the speed of the anaerobic digestion of certain substances and possible inhibitors in the examined concentrations of these substances can be conducted. The results are presented in fig. 1, including a reference curve with corn.

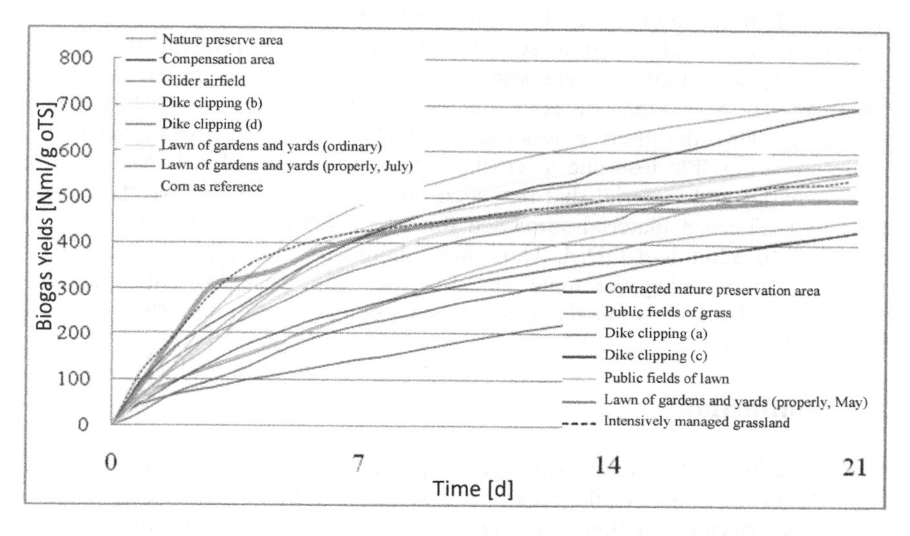

Figure 1: Biogas yields of different grass samples.

The gas yield of the silage of intensively managed grassland is 545 ml_N/g ODM and therefore is a little bit lower than the yields of the outgrowths of nature preserve areas and of levee areas of the first time of cutting it from the end of May to the beginning of July. Since this sample was one of the late clippings of September, the latter were expected to produce less gas. These values are confirmed by literature [12]. The gas yields of extensively cultivated meadows ranged from 325 ml_N/g ODM for silages of the compensation areas up to 705 ml_N/g ODM for dike clippings (c). For these experiments the gas yields of silages are clearly below the yield of the not conserved outgrowths. The gas production of the silage of the compensation areas and the contracted nature preserve areas (325 ml_N/g ODM and 445 ml_N/g ODM) is less than the gas production of the second dry clipping of the glider airfield with 460 ml_N/g ODM. Except for the outgrowths of public fields of grass the substrates of extensive meadows do not show any signs of being hemmed during the gas production. The curve of the gas production shows an obvious sharp bend at the beginning. This might be due to both the high content of nitrogen (2.4%) and an eventual hemming because of ammonia. The gas yield of lawn substrates is in the range of the extensive meadows except for the clippings from the communal maintenance of the public fields of grass. This clipping has an increase in the gas production rate of 720 ml_N/g ODM. Ordinary maintained grass of gardens and yards has a production rate of 530 ml_N/g ODM, as the properly maintained grass with 525 ml_N/g ODM for the clipping of May and 570 ml_N/g ODM for the material of June. The C:N-ratio of the grass substrates are in the optimum range compared to the extensive meadows. Despite the good expectations from this concerning the digestibility of the grass, compared to extensive meadows, the experiments show similar levels of degradability of the organic substance of 64.2% for proper maintained grass and of 71.6% for the public fields of grass.

Concerning literature no losses were predicted, but problems during the ensiling of late clippings were known, since the large structures cannot be densified. The gas yield of both silages, which had 325 ml_N/g ODM and 445 ml_N/g ODM, definitely is less than the yields of not conserved outgrowths. Therefore a possible loss due to ensilaging has to be considered. Additionally fresh material and material that has been dried on the dike for two days were compared. The wet material surpassed the dry material with 705 ml_N/g ODM compared to 520 ml_N/g ODM. Thus there should be a decrease of the biogas potential when drying the material. This was an expected result, since the elemental analysis showed a decrease of the carbon content of 45.99% to 45.63% during dehydration.

5 Combustion

The utilization of the outgrowths can be processed using one of two possible methods. Fresh grass or silage can be digested in a biogas plant or it can be combusted. Because of that a reference is analyzed to determine the calorific value of the gas potential of the samples. The calorific value H_U is defined to be the maximum usable amount of heat during combustion (DIN EN 14918).

Utilizing the grass and lawn clippings is only possible to a certain extent, since some of the properties of the grass stalk are disadvantageous to the combustion process. Sophisticated firing methods and an extended off-gas treatment require much more effort. On the one hand this leads to the exhaustion of more chloral, nitrogen, potassium or sulphuric gases, on the other hand there are increased amounts of ashes, causing higher rates of pollutant emissions and increasing the costs of the firing system. The amount of ashes of the lawn substrates is about 20%, which is more than the ashes of average grass clippings, which is about 9%. Wood is producing a significantly smaller amount of ash, about 0.5% [11]. Additionally the utilization of recently harvested plants is more difficult, since the material is very moist and would have to be dried before firing it. Moisture contents of more than 20% [13] cause the combustion to be inefficient and drying it would require several additional mechanical operations.

Table 3 shows the average moisture contents and amounts of ashes for the samples from the district of Bergedorf. The moisture content is more than the required 20%. The caloric values are about 9% below the average caloric value of wood which is 18.5 MJ/kg TS [14].

Table 3: Biogas yields, caloric values, moisture contents and amounts of ashes.

Substrate	Theoretic Biogas production	Biogas production	Caloric value H_u	Moisture content	Amount of ashes
Unit	l/kg ODM	l/kg ODM	kJ/kg	% ODM	% TS
Intensively managed grassland		545			
Extensive grassland					
Public meadow	852	500	16,537	k.A.	12.9
Silage from compensation area	812	325			
Glider airfield		460			
Dike clipping (a)	812	705	16,928	56.1	8.4
Dike clipping (c)	789	620	16,409	70.2	8.6
Dike clipping (d)			16,356	38.7	8.3
Nature preserve area	773	445	16,157	64.2	7.8
Fields of grass					
Lawn of gardens and yards (ordinary)	878	530	16,684	82.4	13.2
Lawn of gardens and yards (properly)	726	545	12,047	57.7	40.6
Public fields of lawn	851	720	16,710	60.2	14.8

6 Summary

Because of the increased energetic use of biomass the competition for agricultural land between energy crops and actual feed crops was intensified. In addition the decrease of cattle farming in Germany frees grassland which has to be maintained. An energetic utilization of so far unused or not efficiently used green waste would be an important step for the improvement and there exists a large untapped potential. Based on the results of this paper a comparison of the theoretical and practical biogas potential as well as the combustion can be made.

The reference was given by different literature values which varied in a range of 500 ml_N/g ODM up to 700 ml_N/g ODM. The practical biogas production was determined by laboratory digestion experiments. The theoretical biogas production was determined by an element analysis via the modified Buswell-equation. The theoretical potential was between 800 ml_N/g ODM and 878 ml_N/g ODM and much higher than the results of the laboratory tests. The practical experiments displayed a good degradability and resulted in values between 325 ml_N/g ODM and 720 ml_N/g ODM depending on the material and its storage. Looking at figure 1 a rapid digestion seems to be clear, identifiable by the strong plateau phase. This indicates a good anaerobic degradation of the substrate. After

21 days the gas production was still at about 2.5% of the so far produced gas. Testing for a longer period should result in higher gas yields. In addition, more optimization opportunities under the anaerobic digestion and the silage-process must be found. The energetic utilization by fermentation is the best option, because the carbon cycle can be closed with digestate. Late clippings of autumn and the grass of extensive land feature a low gas yield, so the combustion of these outgrowths could be an alternative method. For the economic energy a lot of factors, i.e. different temperatures during the fermentation, the storage, treatment and collection processes and co-substrates, have to be investigated.

References

[1] Rösch, C., *Vergleich stofflicher und energetischer Wege zur Verwertung von Bio- und Grünabfällen.* In: Wissenschaftliche Berichte FZKA 5857, S. 269, 2005.

[2] Pfadenhauer, J., *Vegetationsökologie.* A script, with 64 tables. 2. expanded edition, Eching near München: IHW-Publisher., 1997.

[3] Prochnow, A., Heiermann, M., Idler, C., Linke, B., Mähnert, P. & Plöchl, M., *Biogas vom Grünland:Potenziale und Erträge*, Leibniz Institute of Agricultural Engineering PODMdam-Bornim, 2007.

[4] Degenbeck, M., Bavarian State Institute for Viticulture and Horticulture [Hrsg.]): *Basiswissen Rasenbau. Anlage und Pflege von Rasenflächen*, In: Deutscher Gartenbau, Heft 4; S. 10-12.

[5] Turf specialist agency (University Hohenheim [Hrsg.]): *Kompetenz für Rasen.* With the collaboration of Hartmut Schneider und Wolfgang Hendle., 2007. https://www.uni-hohenheim.de/rasenfachstelle/ [Accessed 29.09.2010]

[6] Buske, M. (MOVECO GmbH [Hrsg]), *Grünland- Abgerenzung, Definition und Unterteilung*, 2010. http://www.architektenleistungen.de/article /Grünland [Zugriff am 15.10.2010]

[7] Oechsner, H. (Universität Hohenheim [Hrsg.]), *Verfahrenstechnik der Vergärung von Biomasse.* State Institute of Farm Machinery and Construction, in collaboration with the Academy for Nature and Environmental Protection of Baden Wuerttemberg, Stuttgart, 2005.

[8] Freie und Hansestadt Hamburg, Ministry of Urban Development and Environment [Hrsg.]): *Nationalpark und Naturschutzgebiete in Hamburg*, 2010. http://www.hamburg.de/contentblob/202306/data/bsu-nsg-tabelle.pdf [Accessed 14.10.2010]

[9] Hofmann, J. (Regierung Niederbayern. Landshut [Hrsg.]), *Grundlagen der Biogaserzeugung*, Proceedings of the briefing on 13 July 2000 in the government of Lower Bavaria, 2001.

[10] Mähnert, P., *Kinetik der Biogasproduktion aus nachwachsenden Rohstoffen und Gülle*, Dissertation, Humboldt-Universität, Landwirtschaftlich-Gärtnerischen Fakultät, Berlin, 2007.

[11] Hartmann, H., Böhm, T. & Maier, L. (Bavarian State Ministry for Regional Development and the Environment (Stmlu) [Hrsg.]), *Naturbelassene*

biogene Festbrennstoffe. Umweltrelevante Eigenschaften und Einflussmöglichkeiten, München, 2000.

[12] Kaiser, F., Einfluss der stofflichen Zusammensetzung auf die Verdaulichkeit nachwachsender Rohstoffe beim anaeroben Abbau in Biogasreaktoren, Dissertation, Maintained by J. Bauer, Technische Universität München, Scientific Center Weihenstephan for Food, Environment and Land Use, 2007.

[13] Elsäßer, M. (Staatliche Lehr- und Versuchsanstalt für Viehhaltung und Grünlandwirtschaft. Aulendorf [Hsrg.], *Möglichkeiten der Verwendung alternativer Verfahren zur Verwertung von Grünlandmähgut: Verbrennen, Vergären, Kompostieren.* In: Reports on Agriculture, 2003.

[14] FNR e.V. [Hrsg.]: *Leitfaden Bioenergie. Planung, Betrieb und Wirtschaftlichkeit von Bioenergieanlagen*; 4., unchanged edition., 2007.

Bioethanol production from oilseed rape straw hydrolysate by free and immobilised cells of *Saccharomyces cerevisiae*

A. K. Mathew, K. Chaney, M. Crook & A. C. Humphries
Harper Adams University College, UK

Abstract

Oilseed rape (OSR) straw can serve as a low-cost feedstock for bioethanol production. Glucose and other fermentable sugars were extracted from OSR straw using sulfuric acid pre-treatment and enzymatic hydrolysis. Batch fermentation of enzymatic hydrolysate with *Saccharomyces cerevisiae* immobilised in Lentikat® was found to be superior to free cells in terms of bioethanol yield. The maximum bioethanol concentration from free and immobilised cells was 6.73 and 9.45 $g.l^{-1}$, respectively, with corresponding yields of 0.41 and 0.49 g bioethanol. g $glucose^{-1}$.
Keywords: bioethanol, dilute acid pre-treatment, immobilisation, oilseed rape straw.

1 Introduction

In 2007, consumption of liquid fuels in the transportation sector was 46 million barrels per day and is expected to increase by 67 million barrels per day by 2035 [1]. In 2007, the use of liquid fuels was responsible for 38% of global greenhouse gas (GHG) emissions, providing a significant contribution to climate change [1]. The replacement of gasoline (petrol) with bioethanol is encouraged globally as a mechanism to reduce exposure to volatility in the oil market, and minimise the extent to which road transport contributes to global warming. Bioethanol can be produced from two different types of feedstocks: first-generation feedstocks (maize, wheat and sugarcane) and second-generation feedstocks (lignocellulosic materials such as straws, forest residue or any agriculture waste) [2]. Commercial production of bioethanol from first-generation feedstocks is limited by land availability, and concerns regarding the use of land for fuel as opposed to food production. Second-generation bioethanol production from lignocellulosic material is a complex process compared to first-generation feedstocks due to the presence of lignin and hemicellulose. Additional processing steps, referred to as pre-treatment and hydrolysis are essential for extracting sugar from lignocellulosic materials. The pre-treatment process is highly energy intensive and expensive (due to enzyme application during hydrolysis), which means the production of second-generation bioethanol is currently non-competitive to first-generation bioethanol [3]. Consequently bioethanol produced from second-generation feedstocks is the focus of considerable research and development.

Global cultivation of OSR was 31 million ha in 2009. Assuming a straw yield of 1.5–3.0 tonnes per ha [4], the amount of OSR straw produced in 2009 was between 46.5 and 93.0 million tonnes. Assuming a bioethanol yield of 270 l $tonne^{-1}$ of straw (using existing technology) [5], it is predicted that between 12.5 and 25.0 billion liters of bioethanol could have been produced from OSR straw. Currently OSR straw does not have an existing market and is normally ploughed back into field. Hence bioethanol production from OSR straw could add value to existing crops.

Dilute acid pre-treatment is one of the most commonly used pre-treatment techniques for altering the structure of lignocellulosic materials [6]. It mainly breaks the structure of hemicellulose and a small portion of lignin. Dilute acid pre-treatment also leads to the formation of fermentation inhibitors such as acetic acid, hydroxymethylfurfural (HMF) and furfural as a result of sugar degradation [7]. Dilute acid pre-treatment has been widely studied for a range of feedstocks. Jeong *et al.* [8] optimised the dilute acid pre-treatment of OSR straw based on the extent to which hemicellulosic sugars (mainly xylose, mannose and galactose) were extracted. Under optimum pre-treatment conditions (1.76% H_2SO_4, 152.6°C for 21 min) 85.5% of total sugars were recovered from OSR straw. The inhibitors present in the pre-treated hydrolysate were acetic acid (2.94 g l^{-1}), 5-hydroxymethylfurfural (0.04 g l^{-1}) and furfural (0.98 g l^{-1}). Subsequent enzymatic hydrolysis resulted in a digestibility of 95.4% after 72 h, compared to a digestibility of 27.1% for untreated OSR straw. Castro *et al.* [9] optimised the dilute acid pre-treatment of OSR straw using pre-treatment temperatures between 140 and 200°C, pre-treatment times between 0 and 20 min and sulfuric acid concentrations between 0.5 and 2.0% (w/w). A mathematical model was used to predict the pre-treatment conditions that would result in a cellulose conversion efficiency of approximately 100%. The optimum conditions were predicted to be the application of temperature at 200°C for 27 min at an acid concentration of 0.40%. Mathew *et al.* [10] studied the dilute acid pre-treatment of OSR straw based on the concentration of glucose recovered after enzymatic hydrolysis. Under optimum pre-treatment conditions (5% (w/w) biomass loading, 2.5% (w/w) acid concentration and 90 min pre-treatment time) 81% of glucan was converted into glucose after 72 h of enzymatic hydrolysis.

The production of bioethanol using immobilised cells has been well studied. However, previous research has focused on the use of either sugar cane or starch hydrolysate as substrate. The advantages of cell immobilisation over free cell fermentation for bioethanol production include a higher volumetric productivity due to higher cell density, enhanced yield and cell viability for repeated cycles of fermentation [11]. The research presented compares the bioethanol yield and volumetric productivity obtained from the batch fermentation of OSR straw hydrolysate using free and immobilised cells of *S. cerevisiae.*

2 Materials and methods

2.1 Microorganism and media

S. cerevisiae Type I was grown at 30°C and maintained on agar slants at 4°C as described by Liu *et al.* [12]. *S. cerevisiae* was cultivated in 150 ml conical flasks with 50 ml growth medium (Glucose, 5.0; yeast extract, 0.5; peptone, 0.5; K_2HPO_4, 0.1; $MgSO_4$. $7H_2O$, 0.1; expressed in g 100 ml^{-1}) and incubated at 30 °C in a shaking incubator at 150 rpm. After overnight incubation (10^8 cells ml^{-1}) cells were harvested by centrifugation at 4000 rpm for 15 min and resuspended in 10 ml growth medium.

2.2 Sugar extraction from OSR straw

OSR straw was pre-treated using the optimum dilute sulfuric acid pre-treatment conditions determined previously [10]. Following pre-treatment, the slurry was filtered, washed with purite water and the liquid fraction collected. The solid fraction was used for enzymatic hydrolysis using cellulase from *Trichoderma ressei* ATCC 26921 (25 FPU g^{-1} biomass) and β-glucosidase from Aspergillus niger (70 CBU g^{-1} biomass) for 72 h, at 50°C and 5% biomass loading using sodium citrate buffer (pH 4.8). The hydrolysate was filtered using No. 1 Whatman filter paper and liquid fraction was used for fermentation.

2.3 Immobilisation of yeast cells

Lentikat® was obtained from geniaLab (Germany) and prepared for immobilisation according to Bezbradica *et al.* [13] after melting at 90 ± 3°C. *S. cerevisiae* cells were immobilised into Lentikat® by mixing 10 ml of *S. cerevisiae* cell suspension with 40 ml of Lentikat® liquid. The mixture was extruded onto petri dishes through a syringe fitted with a needle (1.25 × 40 mm). The petri dishes were left to dry in a laminar flow cabinet under a downwards vertical airstream at room temperature for approximately 2 h. The Lentikat® discs were stabilised and re-swollen in 100 ml stabilising solution (geniaLab, Germany) for 2 h. Lentikat® immobilised *S. cerevisiae* were allowed to proliferate through overnight incubation in 100 ml growth medium according to Liu *et al.* [12]. Lentikat® was selected as an immobilisation support because of its mechanical strength.

2.4 Fermentation conditions

2.4.1 Bioethanol production from enzymatic hydrolysate using free and immobilised cells of *S. cerevisiae*

Batch fermentation was completed using free and Lentikat® immobilised cells of *S. cerevisiae* in 150 ml sterile conical flasks with 50 ml of fermentation medium (enzymatic hydrolysate with glucose concentration of 16–19 g l^{-1}) for 24 h at 30 ± 3°C and 150 rpm. The fermentation medium was composed of glucose, yeast extract 0.5 g 100 ml^{-1}; peptone, 0.5 g 100 ml^{-1}; K_2HPO_4, 0.1 g 100 ml^{-1} and $MgSO_4.7H_2O$, 0.1 g 100 ml^{-1}. The fermentation medium was inoculated (10% w/v) either with immobilised or free cells. Samples (1 ml) were withdrawn from the fermentation medium after 2, 4, 6, 10 and 24 h, and analysed to determine the glucose and bioethanol concentration. Batch fermentations were completed in triplicate.

2.4.2 Bioethanol production by immobilised cells of *S. cerevisiae* using acid pre-treatment and enzymatic hydrolysate as substrate

Batch fermentation was completed using Lentikat® immobilised cells of *S. cerevisiae* as per the method presented in section 2.4.1, but the fermentation medium was composed of the liquid fraction collected immediately following pre-treatment and the enzymatic hydrolysate. The sugars present in the pre-treated hydrolysate were xylose, 7.46 g l^{-1}; glucose, 1.77 g l^{-1}; galactose, 1.35 g l^{-1}; arabinose, 0.85 g l^{-1}. The glucose concentration of the fermentation medium was adjusted to 23 g l^{-1} using pure glucose.

2.5 Analytical methods

The concentration of bioethanol and glucose present in the fermentation media was analysed using HPLC fitted with a refractive index detector. HPLC analysis was completed according to NREL [14] laboratory analytical procedure.

2.6 Statistical analysis

Statistical analysis was completed using Genstat 13th edition. The effect of immobilisation of *S. cerevisiae* in Lentikat® supports and free cells was analysed separately by using one-way analysis of variance (ANOVA).

3 Results

3.1 Bioethanol production from enzymatic hydrolysate using free and immobilised cells of *S. cerevisiae*

Batch fermentation was completed using glucose extracted from the enzymatic hydrolysis of OSR straw as carbon source. Batch fermentation was completed over a 24 h time period and employed either free or Lentikat® immobilised cells of *S. cerevisiae* (fig. 1). The concentration of bioethanol produced after 24 h of fermentation was approximately 40% higher ($p = 0.021$) for immobilised cells

(8.62 ± 0.31 g l^{-1}) compared to free cells (6.15 ± 0.21 g l^{-1}). This finding is supported by Swain *et al.* [11], who demonstrated a 6.7% increase in bioethanol yield when mahula flowers were fermented with cells immobilised in calcium alginate as opposed to free cells. Glucose was fully consumed within the 24 h fermentation time period; hence a lower bioethanol yield cannot be attributed to a reduced level of glucose uptake by free cells. The conversion of glucose to bioethanol was significantly faster with immobilised cells than with free cells. A maximum bioethanol concentration of 9.45 ± 0.25 g l^{-1} was achieved within 4 h of fermentation when immobilised cells were used. This represents a glucose conversion efficiency of approximately 98% compared to 67% glucose conversion by free cells in the same time period. The final biomass concentration recorded in Swain *et al.* [11] was higher in free cell fermentation, suggesting glucose may have been diverted from bioethanol production in order to support cell growth. It took approximately 24 h for free cells to consume 99% of the initial glucose present in the medium, demonstrating cell immobilisation resulted in significantly enhanced volumetric productivity. The volumetric productivity of bioethanol fermentation using immobilised *S. cerevisiae* cells was 4.12 g l^{-1} h^{-1} compared to 1.69 g l^{-1} h^{-1} for free cells after 2 h. The volumetric productivity reduced to 0.26 g l^{-1} h^{-1} and 0.36 g l^{-1} h^{-1} for free and immobilised *S. cerevisiae* after 24 h of batch fermentation due to a lower concentration of glucose present in the fermentation medium. At the initial stages of batch fermentation (until 4 h), the volumetric productivity of immobilised *S. cerevisiae* cells was found to be approximately two to three times higher than that of free cells. According to Nedović and Willaert [25],

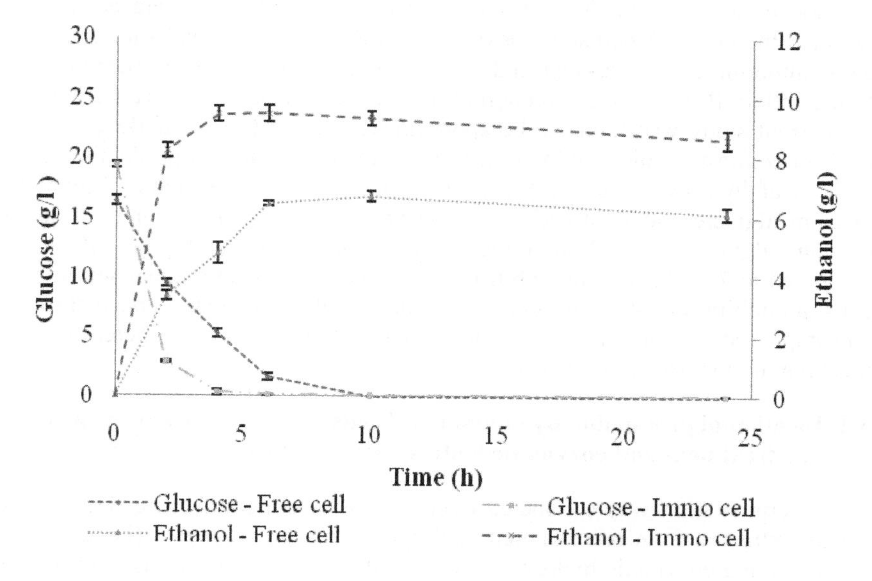

Figure 1: Glucose consumption and ethnaol production using free and immobilised cells of *S. cerevisiae*.

immobilised cells have less hydrodynamic and mechanical stress compared to cells in suspension, which enables the cells to utilise more cellular energy for product formation, resulting an improvement in volumetric productivity.

A decrease in bioethanol concentration was observed with free cells and immobilised cells between 6 and 24 h of fermentation. However statistical analysis suggested this was not a significant reduction ($p = 0.289$ for free cells and $p = 0.224$ for immobilised cells). The same trend was observed in a study conducted by Kuhad *et al.* [15], where the bioethanol concentration and bioethanol yield reduced between 16 and 24 h fermentation. The reduction in bioethanol concentration may be due to the oxidation of bioethanol to acetic acid [16] or due to the formation of other by-products such as glycerol and butyric acid (as a result of contamination) [17]. Another possible explanation is the consumption of accumulated bioethanol by *S. cerevisiae* that has adapted to simultaneously consume fermentable sugars and bioethanol [18]. The maximum bioethanol yield obtained from free cells was 0.41 g bioethanol. g glucose^{-1} and that of immobilised cells was 0.49 g bioethanol. g glucose^{-1}. These yields are less than the stoichiometric bioethanol yield of 0.51 g bioethanol. g glucose^{-1}. The lower bioethanol yield observed may have been due to the use of glucose for cell growth during fermentation, essentially diverting substrate from bioethanol production.

The maximum theoretical bioethanol conversion (%) was found to be 80.51% after 10 h of fermentation with free cells and 95.66% after 4 h of fermentation with immobilised cells. Behera *et al.* [19] studied the production of bioethanol from mahula flowers using *S. cerevisiae* (strain CTCRI) cells either free in solution or immobilised in agar-agar or calcium alginate. Behera *et al.* [19] reported theoretical bioethanol conversions of 87%, 93% and 95% for free cell, cells immobilised in agar-agar and calcium alginate, respectively after 96 h of fermentation. Bioethanol conversion (%) with immobilised *S. cerevisiae* cells in the current study was found to be approximately equal to that of Behera *et al.* [19]. In contrast to this, Rakin *et al.* [20] compared bioethanol production from corn meal hydrolysate using *S. cerevisiae var. ellipsoideus* immobilised into calcium alginate and Lentikat®. *S. cerevisiae var. ellipsoideus* immobilised in calcium alginate resulted in a higher theoretical bioethanol yield of 111% compared to 77% for cell immobilised in Lentikat® discs. In comparison to free cells, immobilised cells were found to result in higher bioethanol concentrations and improved volumetric productivity of batch fermentation by increasing the local population density of cells [21, 22].

3.2 Bioethanol production by immobilised cells of *S. cerevisiae* using acid pre-treatment and enzymatic hydrolysate as substrate

Batch fermentation was conducted over a 24 h time period using fermentable sugars extracted from pre-treatment and hydrolysis of OSR straw (i.e. acid pre-treatment and enzymatic hydrolysis hydrolysate were combined). The results are shown in fig. 2. The bioethanol concentration obtained after 24 h of batch fermentation from acid and enzymatic hydrolysate fermentation medium was

8.66 ± 0.09 g l^{-1}. A maximum bioethanol concentration of 9.26 ± 0.04 g l^{-1} was obtained after 6 h of fermentation. Similarly to when enzymatic hydrolysate was used as sole substrate, there was a significant reduction ($p = 0.044$) in bioethanol concentration between 6 and 24 h of batch fermentation. There was no significant difference observed ($p = 0.873$) in the concentration of bioethanol produced when enzymatic hydrolysate only and acid pre-treatment and enzymatic hydrolysate were used as substrate, even though the initial concentration of fermentable sugars was higher in the latter.

A bioethanol yield of 0.39 g bioethanol g glucose^{-1} was achieved with immobilised cells using acid pre-treatment and enzymatic hydrolysate as substrate. A yield of 0.49 g bioethanol g glucose^{-1} was obtained for immobilised cells when enzymatic hydrolysate was used as substrate. The lower bioethanol yield observed when acid and enzymatic hydrolysate was added to the fermentation medium may have been due to the presence of inhibitors such as acetic acid, hydroxymethylfurfural (HMF) and furfural that are produced during acid pre-treatment and which were present in the pre-treatment hydrolysate. These products can inhibit bioethanol production by reducing the activity of several enzymes such as alcohol dehydrogenase, aldehyde dehydrogenase and pyruvate dehydrogenase [23].

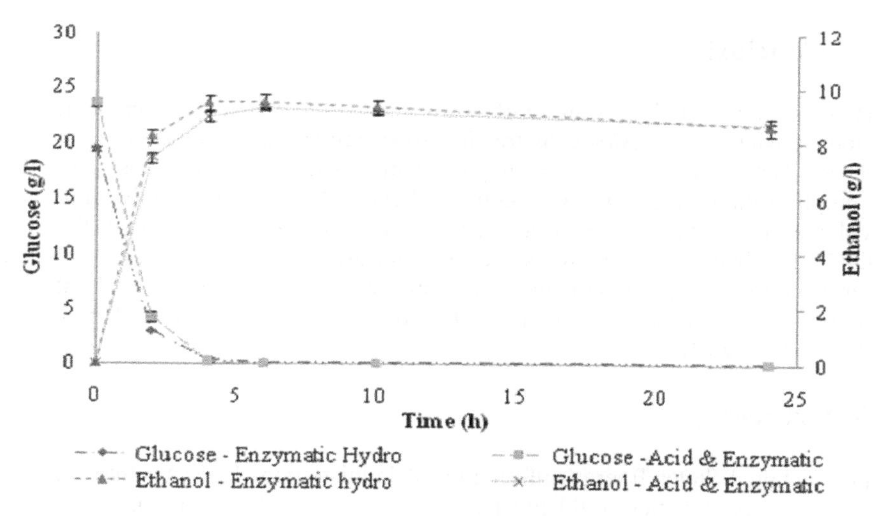

Figure 2: Glucose consumption and bioethanol production from enzymatic hydrolysate, acid and enzymatic hydrolysate using immobilised *S. cerevisiae* cells.

The volumetric bioethanol productivity of immobilised cells when acid and enzymatic hydrolysate was added to the fermentation medium was 2.24 g l^{-1} h^{-1} after 4 h of batch fermentation. After 24 h of fermentation, the volumetric bioethanol productivity was reduced to 0.36 g l^{-1} h^{-1}. This is similar to the volumetric bioethanol productivity observed after 24 h, for free and immobilised cells of *S. cerevisiae* using enzymatic hydrolysate (0.26 and 0.36 g l^{-1} h^{-1}, respectively). Behera *et al.* [19] reported volumetric bioethanol productivities of 0.258, 0.262 and 0.268 g l^{-1} h^{-1}, for cells free in solution, and immobilised in agar-agar and calcium alginate, respectively. The current study demonstrates higher bioethanol productivity in immobilised fermentation compared to Behera *et al.* [19] whereas the volumetric productivity of free cells was found to be similar to the study conducted by Behera *et al.* [19]. Petersson *et al.* [24] demonstrated a bioethanol productivity of 0.91 g $l^{-1}h^{-1}$ from simultaneous saccharification and fermentation of wet oxidised OSR straw. The ethanol productivity in the current study was found to be lower due to a lower concentration of glucose present in the fermentation medium initially.

From the current study, the maximum theoretical bioethanol yield from acid and enzymatic hydrolysate was approximately 76%, which is less than the maximum theoretical bioethanol yield observed from free cells in suspension using enzymatic hydrolysate as fermentation medium. Hence the current study concluded that fermentation using enzymatic hydrolysate would be better than combined use of pre-treatment and enzymatic hydrolysate for bioethanol production. Potentially the bioethanol yield obtained when pre-treatment hydrolysate is used as carbon source could be increased through the fermentation of pentose sugars.

4 Conclusion

Bioethanol production from OSR straw using dilute acid pre-treatment was studied using *S. cerevisiae* cells free in suspension and immobilised in Lentikat® discs. Batch fermentation with enzymatic hydrolysate demonstrated the immobilisation of *S. cerevisiae* cells in Lentikat® discs resulted in improved volumetric productivity and a higher concentration of bioethanol (9.47 ± 0.27 g l^{-1}) than when cells free in suspension were used (6.73 ± 0.18 g l^{-1}). Batch fermentation of pre-treatment and enzymatic hydrolysate using Lentikat® immobilised cells did not improve the concentration of bioethanol produced (9.26 ± 0.04 g l^{-1}).

References

[1] United States Energy Information Administration (USEIA), International energy statistics, 1000 Independence Ave., SW, Washington, DC 20585, http://www.eia.gov/

[2] Balat, M., Production of bioethanol from lignocellulosic materials via biochemical pathway: A review. *Energy Conversion and Management*, **52**, pp. 858–875, 2011.

[3] Lora, E.E.S., Palacio, J.C.E., Rocha, M.H., Reno, M.L.G., Venturini, O.J. & Olmo, O.A. del, Issues to consider, existing tools and constraints in biofuels sustainability assessments. *Energy*, **36(4)**, pp. 2097–2110, 2011.

[4] Newman, R., A trial burn of rape straw and whole crops harvested for energy use to assess efficiency implications, B/U1/00768/00/00

http://webarchive.nationalarchives.gov.uk/+/http://www.berr.gov.uk/files/file14920.pdf

[5] Larson, D.E., Biofuel production technologies: status, prospects and implications for trade and development, united conference on trade and development UNCTAD/DITC/TED/2007/10, http://www.unctad.org/en/docs/ditcted200710_en.pdf

[6] Mosier, N., Wyman, C., Dale, B., Elander, R., Holtzapple, Y.Y.L.M. & Ladisch, M., Features of promising technologies for pre-treatment of lignocellulosic biomass. *Bioresource Technology,* **96**, pp. 673–686, 2005.

[7] Palmqvist, E. & Hahn-Hagerdal, B., Fermentation of lignocellulosic hydrolysates. II:inhibitors and mechanisms of inhibition. *Bioresource Technology*, **74**, pp. 25–33, 2000.

[8] Jeong, T.S., Byung, H.U., Jun-Seok, K. & Kyeong-Keun, O., Optimizing dilute acid pre-treatment of rapeseed straw for extraction of hemicelluloses. *Applied Biochemistry and Biotechnology*, **161**, pp. 22–33, 2010.

[9] Castro, E., Diaz, M.J., Cara, C., Ruiz, E., Romero, I. & Moya M., Dilute acid pre-treatment of rapeseed straw for fermentable sugar generation. *Bioresource Technology*, **102**, pp. 1270–1276, 2011.

[10] Mathew, A.K., Humphries, A.C., Chaney, K. & Crook, M., Dilute acid pre-treatment of oilseed rape straw for bioethanol production. *Renewable Energy*. **36(9)**, pp. 2424–2432, 2011.

[11] Swain, M.R., Kar, S., Sahoo, A.K. & Ray, R.C., Ethanol fermentation of mahula (Madhuca latifolia L.) flowers using free and immobilized yeast *Saccharomyces cerevisiae*. *Microbiology Journal Research*, **162**, pp. 93–98, 2007.

[12] Liu, R., Li, J. & Shen, F., Refining bioethanol from stalk juice of sweet sorghum by immobilised yeast fermentation. *Renewable Energy*, **33**, pp. 1130–1135, 2008.

[13] Bezbradica, D., Obradovic, B., Leskosek-Cukalovic, I., Bugarski, B. & Nedovic, V., Immobilisation of yeast cells in PVA particles for beer production. *Process Chemistry,* **42(9)**, pp. 1348–1351, 2007.

[14] National Renewable Energy Laboratory, http://www.nrel.gov/biomass/

[15] Kuhad, R.C., Gupta, R., Khasa, Y.P. & Singh, A., Bioethanol production from *Lantana camara* (red sage): pre-treatment, saccharification and fermentation. *Bioresource Technology*, **101**, pp. 8348–8354, 2010.

[16] Christensen, C.H., Jorgensen, B., Rass-Hansen, J., Egeblad, K., Madsen, R., Kiltgaard, S.K., Hansen, S.M., Hansen, M.R., Andersen, H.C. & Riisager, A., Formation of acetic acid by aqueous phase oxidation of ethanol with air in the presence of heterogeneous gold catalyst. *Angewandte Chemie International Edition*, **45**, pp. 4648–4651, 2006.

[17] Cheng, J., *Biomass to renewable energy processes*, CRC press, Taylor & Francis group, pp. 240–241, 2009.

[18] Ramon-Portugal, F., Pingaud, H. & Strehaiano, P., Metabolic transition step from ethanol consumption to sugar/ethanol. *Biotechnology Letters*, **26**, pp. 1671–1674, 2004.

[19] Behera, S., Kar, S., Mohanty, R.C. & Ray, R.C., Comparative study of bioethanol production from mahula (*Madhuca latifolia L.*) flowers by *Saccharomyces Cerevisiae* cells immobilised in agar-agar and Ca-alginate matrices. *Applied Energy*, **87**, pp. 96–100, 2010.

[20] Rakin, M., Mojovic, L., Nikolic, S., Vukasinovic, M. & Nedovic, V., Bioethanol production by immobilised *Saccharomyces Cerevisiae var. ellipsoideus* cells. *African Journal of Biotechnology*, **8**, pp. 464–471, 2009.

[21] Yu, J., Zhang, X. & Tan, T., A novel immobilisation method of *Saccharomyces cerevisiae* to sorghum bagasse for ethanol production. *Journal of Biotechnology*, **129**, pp. 415–420, 2007.

[22] Najafpour, G., Younesi, H., Ku, S. & Ku, I., Ethanol fermentation in an immobilised cell reactor using *Saccharomyces cerevisiae. Bioresource Technology*, **92**, pp. 251–260, 2004.

[23] Modig, T., Liden, G. & Taherzadeh, M.J., Inhibition effects of furfural on alcohol dehydrogenase, aldehyde dehydrogenase and pyruvate dehydrogenase. *Biochemical Journal*, **363**, pp. 769–776, 2002.

[24] Petersson, A., Thomsen, M.H., Nielsen, H.H. & Thomsen, A.B., Potential bioethanol and biogas production using lignocellulosic biomass from winter rye, oilseed rape, faba bean. *Biomass and Bioenergy*, **3**, pp. 812–819, 2007.

[25] Nedović, V. & Willaert, R., eds., *Applications of cell immobilisation biotechnology*, Springer: Dordrecht, Netherlands, 2005.

Assessment of the potential of residuary microalgae from stabilization ponds for the production of biofuel

G. E. G. Vieira[1], A. da S. Cardoso[1], A. K. Marques[1] & A. Pickler[2]
[1]*Federal University of Tocantins, Brazil*
[2]*Cenpes/Petrobras, Brazil*

Abstract

Biomass with high lipids is necessary to the biodiesel industry since the demand for this biofuel is increasing in Brazil and the main oilseed crop used for producing biodiesel in Brazil will not be able to aid the production of biodiesel without arable lands destined only for that purpose. Thus, the microalgae emerges as a potential biomass and they are being used as a source of products for biofuel production because of their high productivity and rapid growth. However, the high cost of microalgal biomass production is the main factor that prevents the use of its products on a large scale. Thus, this work is aimed to use the microalgal biomass already present in effluents from aerobic and anaerobic lagoon ponds used for processing the effluent of the UASB reactor to obtain products and it is proposed to use these as a biofuel. Initially, a qualitative and quantitative characterization of the microalgae species present in the aerobic anaerobic lagoon was performed. The biomass was recovered from the effluent using a vacuum filtration system and was sun dried. For the extraction of total lipids, a mixture of polar and apolar solvents was used. The extract was analyzed using FTIR and GC-MS. The results showed that, in quantitative terms, the taxa Cyanobacteria and Euglenophyceae were predominant. In relation to the species found, the species with the largest number of individuals per milliliter were *Lepocinclis salina* Frits., *Plankthotrix isothrix* Bory and *Euglena* sp. The extract obtained presented values of 16 wt%. The infrared spectrum showed the presence of ester and alkanes just like the GC-MS analysis. The microalgae identified are producers, predominantly, of lipids with saturated fatty acids and hydrocarbon. The high presence of microalgae in a natural watercourse can bring environmental problems. Thus, obtaining

microalgal biomass aerobic and anaerobic from a lagoon pond might provide an economic and environmental gain.
Keywords: effluent from aerobic and anaerobic lagoon, lipids, hydrocarbon, biodiesel from microalgae, environmental impacts.

1 Introduction

Biomass with high lipids has been the target of research to ensure the supply of triglycerides for biodiesel plant (simple alkyl esters of fatty acids), since the demand for this biofuel is increasing in Brazil (3.6% per year) and it is observed that the main oilseeds used to obtain biodiesel in Brazil (soybeans, serving 75% of biodiesel production) will not be able to aid the production of this biofuel without the allocation of agricultural lands solely for that purpose, due to the low productivity of biomass (0.2–0.4 tons per hectare). Therefore, it is preferable that the biomasses used provide an optimum productivity in lipids with the use of a smaller surface area of land [1–3].

As a promising source of lipids for the biodiesel production chain, microalgae appear as a biomass that have high quantities of lipid (20%–50%), which is obtained naturally from the conversion of atmospheric carbon dioxide by the use of photosynthesis and assisting in the removal of CO_2 from the atmosphere [1]. These lipids are composed of glycerol, sugars or esters of saturated or unsaturated fatty acids (12 to 22 carbon atoms) [4], with its content and composition influenced by factors such as light, temperature, concentration and source of nitrogen and concentration of CO_2 [5] and can be obtained by mechanical (press) and/or the use of solvents, as it is done with traditional biomass [1]. What is observed is that the lipid content of microalgae and their productivity is much higher than some oilseed traditionally used to produce biodiesel.

The microalgal biomass is mainly obtained from cultivation in the laboratory or industrial closed systems (photobioreactors), or aerated ponds fed with culture medium [6]. In these, there is a predominance of one species by the use of microalgae cultivated by autotrophic and/or heterotrophic processes. In a way to minimize the costs of the final product (microalgae oil), microalgae can be obtained directly from a system where they are already available naturally, due to the provision of optimal environmental conditions for its development. The microalgae fed by effluent in a stabilization pond system, for example, can be used to obtain cheaper oil when compared to those obtained by a cropping system [7]. This will lead to saving water and nutrients as well as serve to remove nutrients from wastewater, such as nitrogen and phosphorus.

The aerobic and anaerobic lagoon stabilization pond is rich in nutrients; in the aerobic zone in a pond of this system is the development of a species of microalgae. These, when performing the process of photosynthesis, provide oxygen for respiration of the bacteria present in this system [8, 9]. In addition, the effluent from the pond with high microlagal biomass concentration is usually released into water, and it may impact on the characteristics of areas fed by the water [9, 10]. Therefore, this medium presents itself as a potential supplier of microalgal biomass.

The high cost of oil from microalgae is the main factor that hinders its widespread use. This is related to spending on infrastructure (construction of tanks and photobioreactors, for example) for cultivation, use of media, and separation processes of microalgae from the medium liquid. Thus, in order to contribute to reducing the value of the oil from the microalgae and remove the large amount of biomass released into natural water, after treatment of the effluent in aerobic and anaerobic lagoon, the intention here is to use the microalgal biomass (called residual microalgae) already present in the effluent from aerobic and anaerobic lagoon ponds of the Wastewater Treatment Plant (WWTP) of Union Village, in Palmas-TO, to obtain lipids and propose the use for biodiesel production.

2 Methodology

2.1 Characteristics of the WWTP Union Village, Palmas-TO

The WWTP Union Village is located in the northern sector of the city of Palmas, state of Tocantins and belongs to the Sanitation Company of the state of Tocantins, consisting of pre-treatment (grating, box of sand and grease trap), an upflow anaerobic sludge blanket (UASB) reactor followed by an aerobic and anaerobic pond (length: 256 m, width: 110 m, area: 28.160 m^2; volume: 42.240 m^3; height: 1.5 m) for a post-treatment of that effluent. The aquatic environment receiving the effluent from the WWTP is the Stream Agua Fria whose mouth is the reservoir of Luis Eduardo Magalhães Hydroelectric Plant, where the release point of the effluent is located near the mouth of the stream. The population makes use of it for various purposes (recreation, fishing and doing their laundry) [11].

2.2 Collection of microalgae

For the qualitative and quantitative analysis of microalgae, an aerobic and anaerobic lagoon sample was collected from the effluent (100 ml) in an amber bottle with the aid of a collecting vessel and a funnel, both made of plastic. The sample was stored and transferred for analysis. Then, a volume of 10 l was collected to obtain biomass. The collection was gathered at the outlet of the pond. After collecting the effluent, it was transported to a laboratory to be made into the concentration of microalgal biomass.

2.3 Qualitative characterization of the microalgae from the aerobic and anaerobic pond

The methodology used for the qualitative analysis of the microalgae from the aerobic and anaerobic pond was described by Marques [12]; this consists of the taxonomic approach and the population structure, using optical microscopy (using a 40x objective, corresponding to the range from 10 µm to 25.30 cm) binocular Olympus MX41, equipped with digital color camera and ocular micrometer; the organisms were observed in frontal, apical and lateral view, aiming to identify, measure and capture images. The characteristics (morphological and morphometric of the vegetative and reproductive life) of taxonomic value of the species were analyzed according to specialized bibliographies.

2.4 Quantitative analysis of the microalgae from the aerobic and anaerobic lagoon

Quantitative analysis was performed according to the method described by Uthermöl [13], which is based on the random distribution of individuals in the bottom of the sedimentation chamber. The supernatant volume was 10 ml, in function of the sample concentration. The count was performed on an inverted microscope whose counted fields were distributed in parallel vertical transects, covering almost the entire area of the chamber, drawn randomly. The sedimentation time was at least 3 hours for each inch of height of the camera [14]. In calculation of the phytoplankton, it was considered as an individual unicellular organism, filaments, trichomes, colognes and cenobio. The number of fields counted was the necessary amount to reach 100 individuals of the more frequent species, or stabilizing the number of species added per field (minimum area of compensation). Lund *et al.* [15] considered the counting error less than 20% at a significance level of 95%. The count results were expressed as individuals per ml, calculated according to Nogueira and Rodrigues [16]. Both the quantitative and taxonomic analyses were performed in duplicate.

2.5 Harvest of microalgal biomass

The concentration of microalgal biomass was carried out using a vacuum filtration system which consisted of a Kitassato, Büchner funnel and vacuum pump (Prismatec model 131B) using, for the retention of microalgae, a quantitative paper filter (CAAL N°1540). After the filtration, the concentrated biomass was dried in direct sunlight, open air (outdoors), and temperature and humidity were measured using a TFA thermo-hygrometer. The concentration of microalgal biomass was performed in triplicate.

2.6 Lipids extraction

The extraction of lipids from the biomass were performed according to Folch *et al.*, as described by Lourenço [4], with adaptations. Initially, the dried samples (343.07 ± 5.44 mg) were put into a beaker and to this, 60 ml of a chloroform:methanol (2:1 v/v) mixture was added. The mixture was stirred (with the aid of mechanical stirrer Fisatom model 713D) in the chapel, and macerated for 5 minutes. Then, a vacuum filtration was made, where the cell debris remained in the filter paper and the medium liquid was collected, its volume was measured and transferred to a separated funnel. To this was added a solution of KCl 0.88% (corresponding to ¼ of the volume of the mixture contained in the funnel). The mixture was stirred manually three times and after this, the medium was put to sit for 10 minutes. The separation was made of the phases (top and bottom), and the upper phase was discarded. The lower phase, dense and with dissolved lipids, was separated and added to a mixture of methanol:distilled water (1:1 v/v) corresponding to ¼ of the solution volume. Again the system was agitated three times and left to stand for 10 minutes. The lower phase was separated and dried with anhydrous Na_2SO_4. Then the mixture was filtered in a vacuum. The solvent present in the mixture was removed by rota-evaporation using a rota-evaporation apparatus Fisatom model 802. The concentrate was measured and its mass was stored in an amber bottle and refrigerated.

2.7 Analysis of the extract by Fourier transform infrared (FTIR) and gas chromatography with mass spectroscopy (GC-MS)

The infrared spectra covering the region of 4000–400 cm^{-1} were obtained in FTIR Spectrophotometer, Thermo Nicolet model Nexus 470, from the analysis of 1 ml of sample extract obtained in the previous step. The spectra were obtained at room temperature in solid pellets, and were acquired with a resolution of 4 cm^{-1} and 32 scans min^{-1}.

2.8 Obtaining and analysis of biodiesel

To obtain the biodiesel 180.4 mg of lipid extracted from the microalgae collected in the WWTP pond were used. The methodology used was described by Xu *et al.* [17]. The biodiesel was analyzed by FTIR, as it was done for the lipid extract. The chromatographic analysis was performed according to Xu *et al.* [17], using equipment of GC-MS Varian Saturn 2200.

3 Results and discussions

3.1 Qualitative characterization of the microalgae from the aerobic and anaerobic lagoon

Descriptive species of the phytoplankton of the aerobic and anaerobic lagoon belong to the groups Bacillariophyceae, Cyanobacteria, Chlorophyceae, Cryptophyceae, Euglenophyceae, Dinophyceae and Zygnemaphyceae. Two taxa were identified belonging to Bacillariophyceae division, 9 taxa of Cyanobacteria division, 9 taxa of Chlorophyceae division, one of the Cryptophyceae, 9 of Euglenophyceae, 7 Zygnemaphyceae and 1 taxon of Dinophyceae division.

3.2 Quantitative analysis of the microalgae from the aerobic and anaerobic lagoon

The largest number of individuals per milliliter was microalgae belonging to Cyanobacteria and Euglenophyceae groups in the two collections. This may be related, according to Soldatelli [18], to the characteristics of the culture medium (secondarily treated effluent by anaerobic UASB), rich in nitrogen compounds, enabling the environment for the proliferation of these groups of microalgae.

Fifteen taxa were identified at generic, specific and intraspecific levels. The genera numbers for different groups of algae were: 3 Cyanobacteria, 5 Chlorophyceae, 1 of Cryptophyceae, 5 Dinophyceae and 1 Euglenophyceae. In relation to the species found, the *Lepocinclis salina* Frits. the *Plankthotrix isothrix* Bory, *Euglena* sp. and *Phacus longicauda* (Ehr.) Duj. were the species with the highest number of individuals per milliliter.

3.3 Harvest of microalgal biomass

The amount of microalgal biomass obtained by filtration of the effluent from the aerobic and anaerobic pond was 821.77 ± 266.40 mg. The effluent, after filtrated, had its apparent color changed (from green to transparent), as shown in fig. 1, confirming the recovery of microalgal biomass and improving the effluent to be released in the receiving body, since, according to Petry [19], when the receiving body has a combination of algae and halogenated compounds, trihalomethanes can be produced. Borges, according to Petry [19], analyzed samples of water containing chlorophyceae algae and cyanobacteria and noted that trihalomethanes were formed in these waters, even in small quantities compared with waters rich in humic acids. He noted also, by filtration or not of water containing algae, that the algae as well as its extracellular products act as precursors of trihalomethanes, where the water with algae in suspension has a higher risk for formation of these byproducts. According to Naval *et al.* [11], the WWTP Union Village generates a lot of negative impacts on various environments (water, soil and air) and the population that uses the stream water that receives the effluent from the WWTP. With respect to the elevated presence of microalgae in the effluent of the WWTP, these authors show that, when thrown into the stream, algal biomass might generate impacts arising from the presence of toxin producing microalgae, might lead to a reduction of the photic zone of the stream and might alter the aquatic flora.

Figure 1: Effluent from the aerobic and anaerobic pond of the WWTP Union Village. Before (a) and after (b) filtration.

3.4 Lipids extraction

In percentage terms, an extract with values of 16wt% was obtained, with an average of 55.6 ± 1.044 mg. Thus, considering only 0.50 m deep of the aerobic and anaerobic lagoon (area of greater light intensity, according to Von Sperling [9]), the volume of 14,080 m^3 would be obtained and, using the least amount of biomass obtained (594.2 mg), recovered algal biomass would be approximately 858.086 Kg and an amount, considering a 16% content, of approximately 127.39 Kg in lipid.

The concentrate obtained had a greenish color and a waxy appearance, as seen in fig. 2. According to Thompson, Jr. [20], some microalgae have an alternative form of lipids storage. The species *Euglena gracilis* (Euglenophyceae), depending on the environmental condition of light in which is found, can store wax composed of fatty acids from 12 to 13 carbon atoms, with a greater quantity of wax when the cells are grown in an anaerobic environment. Besides wax, many species identified in the effluent of the aerobic and anaerobic pond in Union Village are producing saturated fatty acids, providing a solid consistency to the extract obtained, when present in the triglyceride structure.

Figure 2: Lipid extract obtained of microalgae from the WWTP Union Village pond. Concentrate (a) and diluted in hexane (b).

3.5 Analysis of the extract by Fourier transform infrared (FTIR)

According to Yang *et al.* [21], most of the absorption bands representing the functional groups of triglycerides can be observed around 2937 cm^{-1} (C-H asymmetric axial deformation), 2856 cm^{-1} (axial symmetric deformation of C-H), 1749 cm^{-1} (axial deformation of C=O group), 1454 cm^{-1} (angular deformation such as scissoring of C-H), 1166 cm^{-1} (angular deformation of C-O and C-H) and 709 cm^{-1} (angular deformation such as rocking of C-H).

Several absorption bands were obtained from the FTIR analysis of the obtained extract, as shown in fig. 3. The numbers 1, 2 and 3 highlight the absorption bands characterizing the presence of esters [22].

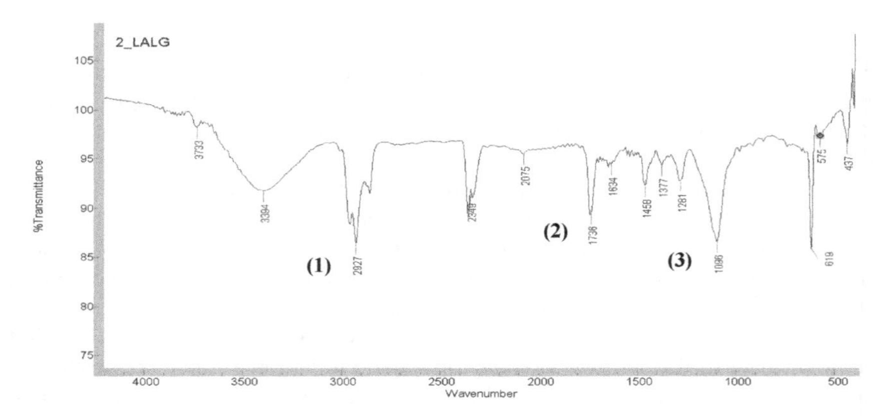

Figure 3: Infrared spectrum of the extract obtained from the microalgae of the WWTP Union Village and diluted in hexane. Axial deformation of C-H of alkanes (1) (2927 cm^{-1}), axial deformation of C=O of saturated aliphatic esters (2) (1736 cm^{-1}), axial deformation of O-C-C of esters (3) (1096 cm^{-1}).

3.6 Analysis of the biodiesel by Fourier transform infrared (FTIR) and gas chromatography with mass spectroscopy (GC-MS)

Figure 4 shows the FTIR spectrum of the lipid sample obtained from the microalgae collected in the pond and transesterified.

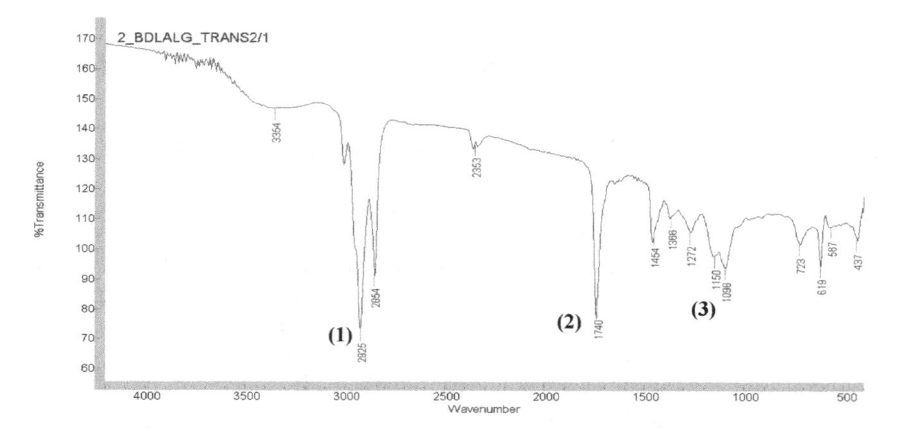

Figure 4: Infrared spectrum of the biodiesel from extract. Axial deformation of C-H of alkanes (1) (2925 cm^{-1}), axial deformation of C=O of saturated aliphatic esters (2) (1740 cm^{-1}), axial deformation of O-C-C of esters (3) (1150 cm^{-1}).

Several absorption bands were obtained from the FTIR analysis of the obtained biodiesel, as shown in fig. 4. The numbers (1), (2) and (3) highlight the absorption bands characterizing the presence of esters [22]. The spectrum of the biodiesel sample from the microalgae obtained lipid showed some similarities with the spectrum of this lipid, but with differences such as the reduction of the O-H group band at 3354 cm^{-1}.

The compounds observed by the analysis of GC-MS were Eicosane ($C_{20}H_{42}$) (42.801%) (1), Hexadecanoic acid methyl ester ($C_{17}H_{34}O_2$) (8.831%) (2), 12.15-Octadecadienoic acid, methyl ester ($C_{19}H_{34}O_2$) (8.803%) (3), 6-Octadecenoic acid, methy ester ($C_{19}H_{36}O_2$) (15.679%) (4), Hexanedioic acid, bis(2-ethylhexyl) ester ($C_{22}H_{42}O_4$) (10.004%) (5) and 1,2-Benzenedicarboxylic, diisooctyl ester ($C_{24}H_{38}O_4$) (13.852%) (6), as shown in fig. 5.

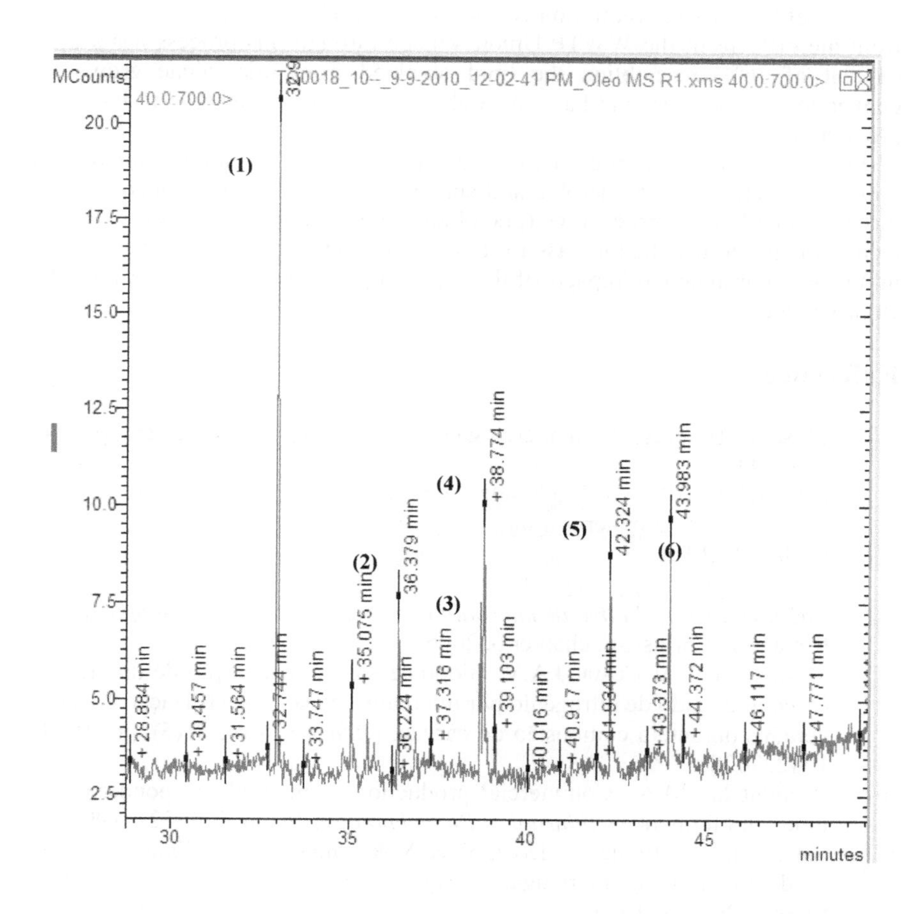

Figure 5: Spectrum of GC-MS of the extract from the microalgae after being derivatized.

4 Conclusions

The effluent of the aerobic and anaerobic lagoon of the WWTP Union Village is composed of different species. The taxa Cyanobacteria, Chlorophyceae, Euglenophyceae and Zygnemaphyceae present the highest number of species. However, with regard to the number of individuals per milliliter, species belonging to taxa Cyanobacteria and Euglenophyceae are predominant. The identified microalgae are producing lipids and fatty acids, which can be used to obtain biodiesel.

In the recovery of microalgal biomass, the filtration process can be used for the microalgal biomass concentration considering a bench scale. Biomass recovered from the effluent of the WWTP Union Village through this process has a lipid content of 16%, resembling the lipid obtained from microalgae grown by conventional processes, and has potential for use in obtaining lipid to biodiesel production.

From an economic and environmental standpoint, obtaining microalgal biomass directly from the aerobic and anaerobic lagoon shows an alternative to the traditionally used procedures (photobioreactors and aerated tanks) to obtain algal biomass, to use the biomass for biofuel production and, simultaneously, to minimize environmental impacts of the elevated presence of this material on the environment.

References

[1] Chist, Y., Biodiesel from microalgae. *Biotechnology Advances*, **25**, pp. 294–306, 2007.

[2] BIODIESELBR,www.biodieselbr.com/noticias/biodiesel/bndes-enxerga-algas-alternativa-materia-prima--grande-potencial-.htm

[3] BIODIESELBR, www.biodieselbr.com/noticias/biodiesel/gargalos-biodiesel-preocupam-bndes.htm

[4] Lourenço, S.O., *Cultivo de microalgas marinhas – princípios e aplicações*. Rima: São Carlos-SP, chapter 6, 2006.

[5] Morais, M.G. & Costa, J.A.V. Bioprocessos para remoção de dióxido de carbono e óxido de nitrogênio por microalgas visando a utilização de gases gerados durante a combustão do carvão. *Química Nova*, **31(5)**, pp. 1038–1042, 2008.

[6] Borowitzka, M.A., Commercial production of microalgae: ponds, tanks, tubes and fermenters. *Journal of Biotechnology*, **70**, pp. 313–321, 1999.

[7] Sawayama, S.; Inoue, S.; Dote, Y. & Yokoyama, S., CO_2 fixation and oil production through microalga. *Energy Conversion and Management*, **36(6-9)**, pp. 729–731, 1995.

[8] Craggs, R.J.; McAuley, P.J. & Smith, V.J. Wastewater nutrient removal by marine microalgae grown on a corrugated raceway. *Water Resource*, **31(7)**, pp. 1701–1707, 1997.

[9] Von Sperling, M., *Lagoas de Estabilização.* UFMG. Departamento de Engenharia Sanitária e Ambiental. Princípios do tratamento biológico de águas residuárias: Belo Horizonte – MG, v. 3, 2.ed. 1986.

[10] Von Sperling, M., *Introdução à qualidade das águas e ao tratamento de esgotos.* UFMG. Departamento de Engenharia Sanitária e Ambiental. Princípios do tratamento biológico de águas residuárias: Belo Horizonte – MG, v. 1, 3.ed. 2005.

[11] Naval, L.P.; Barros, D.S.B. & Silva, C.D.F., Aplicação de métodos de avaliação de impactos ambientais a uma ETE operando em escala real com a respectiva proposição de medidas mitigadoras. www.bvsde.paho.org /bvsAIDIS/PuertoRico29/danger.pdf

[12] Marques, A.K., Análise da diversidade fitoplanctônica no reservatório da Usina Hidrelétrica Luís Eduardo Magalhães, no médio Tocantins – TO: estrutura da comunidade, flutuações temporais e espaciais. www.site.uft.edu.br/index.php?option=com_docman&task=doc_details&g id=974&Itemid=69

[13] Uthermöl, H., Zur Vervollkomninung der quantitativen Phytoplankton. *Methodik Mitteilungen Internationalen Vereinigung fur Theoretische und Angewandte Limnologie*, **9**, pp. 1–38, 1958.

[14] Margalef, R., *Limnologia*, Omega S/A: Barcelona, 1010 p. 1983.

[15] Lund, J.W.G.; Kipling, C. & Lecren, E.D., The inverted microscope method of estimating algal number and the statistical basis of estimating by counting. *Hydrobiologia*, **11**, pp. 143–170, 1958.

[16] Nogueira, I.S. & Rodrigues, L.N.C., Planctonic algae of an artificial lake of Chico Mendes botanical gardens, municipality of Goiania, state of Goias: floristic and some ecological considerations. *Revista Brasileira de Biologia*, **16**, pp. 155–174, 1999.

[17] Xu, H.; Miao, X. & Wu, Q., High quality biodiesel production from a microalga *Chlorella protothecoides* by heterotrophic growth in fermenters. *Journal of Biotechnology*, **126**, pp. 499–507, 2006.

[18] Soldatelli, V.F., Avaliação da comunidade fitoplanctônica em lagoas de estabilização utilizada em tratamento de efluentes domésticos. Estudo de caso – ETE – UCS. www.lume.ufrgs.br/handle/10183/10953

[19] Petry, A.T., Efeito potencial de gradiente trófico em rio urbano na formação de trihalometanos. www.dominiopublico.gov.br/pesquisa/ DetalheObraForm.do?select_action=&co_obra=117560

[20] Thompson, Jr., G.A., Lipids and membrane function in green algae. *Biochimica et Biophysica Acta*, **1302**, pp. 17–45, 1996.

[21] Yang, H.; Irudayaraj, J. & Paradkar, M.M., Discriminant analysis of edible oils and fats by FTIR, FT-NIR and FT-Raman spectroscopy. *Food Chemistry*, **93**, pp. 25–32, 2005.

[22] Silverstein, R.M. & Webster, F.X., *Identificação espectrométrica de compostos orgânicos*, LTC – Livros Técnicos e Científicos Editora S.A: Rio de Janeiro – RJ, cap. 3, 2000.

Biomass pellet production with industrial and agro-industrial waste

J. C. A. R. Claro & D. Costa-Gonzalez
Universidade de Trás-os-Montes e Alto Douro, Portugal

Abstract

Based on a patented process for the treatment and reprocessing of waste and effluents from olive oil and cork industries (such as two-phase olive husks cork powder), new research has been carried out incorporating other industrial waste. In previous research, using only olive oil and cork waste, biomass pellets were obtained having high calorific values (20.6 MJ/kg and 22.2 MJ/kg), low moisture (8.9%) and relatively a low ash percentage (3.9%). Several new formulations and their relevance to the calorific value were now tested using other waste or by-products. The potential inclusion of olive leaves from the olive oil industry, grape husks and grape seeds from the winery industry and chestnut shells (outer shell and inner skin) from the chestnut industry has been analysed. This process will contribute to solving environmental problems resulting from the discharge of those waste and effluents and to create a viable and profitable alternative to their storage and/or deposition in landfill. The increase of other biomass sources for these types of process is extremely important from an environmental and economic management point of view. This research shows that the use of different types of raw materials as biomass sources (giving different pellets formulations) results in a different final product with different physical and chemical characteristics that may improve its quality.
Keywords: biomass pellets, olive waste, cork waste, food industry waste.

1 Introduction

Company strategies are now becoming increasingly focused on factors for increasing productivity which also profess to be “environmentally-friendly”. The reprocessing of waste and by-products is therefore at the top of corporate concerns.

Given the strategic importance that the olive oil sector represents for Mediterranean countries, these production and transformation units must not disregard this new attitude.

In this context, this process could be important to biomass industries and constitutes a very important technological platform internally for olive oil companies, launching them into a strategic framework characterised by an increase in productivity, a cleaner and more environmentally-friendly production that will become an important contribution to the sustainability of the sector.

Olive oil can be obtained by three processes each of which includes the phases of weighing, washing of the olives, warehousing and milling (or grinding). The traditional process involves a pressing phase of the olives followed by a decantation/centrifugation which produces husks and olive mill wastewater in addition to the olive oil.

More modern methods include, a "beating" process replacing the pressing, followed by an "extraction" in horizontal centrifuges. Water is added during these last two phases to facilitate the separation of the olive oil. These more modern methods consist of:

- Phase 1: Here the effluents are the same as those in the traditional method (husks and olive mill wastewater which are separate);
- Phase 2: This process results in a single effluent, known in the industry as "humid husks" or a "paste" (the result of the husks mixed together with the olive mill wastewater).

Most of the olive oil currently produced uses the two production processes above. There are also production units known as "husk oil extraction units" which extract some of the remaining olive oil from the "husks" or "humid husks".

In any case and irrespective of the method used, all the units generate waste and effluents which are harmful to the environment [1–3] and extensive research has been carried out on this issue [4–6].

Some existing treatment and/or reprocessing systems for effluents from olive oil production units are: irrigation of agricultural soils, lagooning, concentration through evaporation, physical/chemical processes, thermal processes, biological processes.

On the other hand, as far as the cork industry residues are concerned, (mainly cork powder, particles less than 0.25 mm in size), which are considered as industrial residues (Code 03 01 99 of European Waste List), problems associated with drainage and storage have been verified, as well as the harmful environmental effects caused by them. The study by the Industrial Association of the District of Aveiro ("Multi-Sector Study on the Area of Environment") in 2000 must be highlighted here (AIDA [7]). This study explicitly refers to the fact that "The production of cork powder is, inclusively, responsible for some physiographical changes verified in the Council of Santa Maria da Feira (small valleys that disappear due to the continual deposition of cork dust in them)". However, the cork powder has had its main use as a combustible fuel for producing energy (burns in kilns), with a small fraction being used for filling in corks of a lower quality, in the linoleum factory and in the control of soils (Gil [8]).

Therefore it can be seen that none of the processes already known represent a sole and universal solution for the effective treatment and reprocessing of waste and effluents from olive oil production units and which at the same time is a solution for the drainage and efficient reprocessing of cork industry waste.

The process presented in this paper can be applied to the treatment of effluents and waste from olive oil production units in the following situations:

- Through the separate treatment of the waste and the effluent (husks and olive mill wastewater);
- Through the joint treatment of the waste and the effluent (humid husks).

In this sense the process can be applied to any type of olive oil production installation whether it is using the traditional production system or the modern two-phase or three-phase production system. Therefore this approach could be an important contribution to a complete and universal solution for the efficient treatment of such residues and effluents. In fact, the known technologies only present part of the solution to the problem and some infer high implementation costs.

2 Methods and materials

2.1 Sample preparation

In sample preparation, two-phase olive husks and industrial waste or by-products, such as olive leaves, grape husks, grape seeds and chestnut shells are used; these were previously turned into powder, in a grinding process using a sieve with a 1 mm mesh.

The powder, which results from the solid materials, is mixed mechanically with the two-phase olive husks generating a pulp material or paste that later is extruded and passes through a drying system to obtain a dried solid product.

2.2 Tested formulations

The olive mill waste (two-phase olive husks) were mixed with the different industrial waste, testing two formulations with 8 and 15% (wt/wt) of each industrial waste powder.

2.3 Gross Calorific Value determination

The Gross Calorific Value (GCV) was determined for each formulation by calorimetric analysis performed with an AC600SHC Semi-Automatic Calorimeter (Leco) in the Chemistry Department of Trás-os-Montes e Alto Douro University (UTAD) in Vila Real, Portugal.

3 Results and discussion

The process utilises grinded industrial waste, namely olive leaves from the olive oil industry, grape husks and grape seeds from the winery industry and chestnut shells (outer shell and inner skin) from the chestnut industry, mixed mechanically with the effluents and/or waste from olive oil production units creating a pulp material or paste that can be used, after drying, as a source of energy fuel. After the mixing, the paste can be extruded in the required form (e.g. pellets or briquettes) and brought to convenient moisture in a dryer. The obtained dried solid product (fig. 1) has a calorific value about 20% higher than wood pellets or chips and a low percentage of moisture and ashes. With this very high calorific value, the recovery of the dry product as pellets, chips, briquettes or logs for burning in a biomass boiler will be extremely feasible.

Figure 1: Biomass samples obtained by the described process.

In table 1, the results of the gross calorific value for the different formulations of the biomass pellets obtained by this process can be seen.

These results show that the use of different types of raw materials as biomass sources for pellet production could be adopted based on this process and a similar range of calorific values are obtained (21.3–23.0 MJ/kg) comparatively to those with the cork powder. Different formulations lead to different final products with different physical and chemical characteristics that may define and improve its quality. Additionally the pellets biomass does not release unpleasant odours and has good mechanical resistance. Nevertheless, more research is needed to design pellets with suitable properties for specific applications and achieve all standards and norms required in the market.

4 Conclusions

This approach offers an important contribution to a complete solution for the residues and/or effluents of the olive oil production units and the end-product obtained does not create a new environmental problem. The final product of this process has a high calorific value, about 20% higher relatively to wood pellets or chips and constitutes a material with great potential as biomass fuel. This process also presents a solution to an environmental problem which is duly identified and regulated by European standards and national laws.

Table 1: Gross Calorific Value (GCV) from biomass samples of olive mill waste (two-phase olive husks) with 8 and 15% (wt/wt) of different kinds of industry waste.

Industry biomass waste	Gross Calorific Value (GCV) MJ kg^{-1}		
	Sample nº	8% (wt/wt)	15% (wt/wt)
Olive leaves	1	22,5390	22,0738
	2	22,4635	22,2682
	3	22,6383	22,1495
	Average	**22,5469**	**22,1638**
	S.D.*	0,0609	0,0696
Grape husks	1	22,8858	23,0456
	2	22,8237	22,9319
	3	22,7819	22,7606
	Average	**22,8305**	**22,9127**
	S.D.*	0,0369	0,1014
Grape seeds	1	22,8982	23,0718
	2	23,0390	23,1672
	3	22,8415	22,8296
	Average	**22,9262**	**23,0229**
	S.D.*	0,0752	0,1288
Chestnut shell	1	21,9291	21,2779
	2	22,1941	21,3670
	3	22,0296	21,3221
	Average	**22,0509**	**21,3223**
	S.D.*	0,0954	0,0298

*S.D. = Standard Deviation.

This enables the difficulties and disadvantages of the existing technologies to be overcome, especially those related to high implementation/execution costs and to the fact that they do not translate into a global solution, contributing only partial and/or one-off solutions to the problem.

In addition to all the environmental benefits, the process enables a product that has a commercial value to be obtained and which may constitute an attractive financial compensation for the olive oil production units and for the cork transformation units. Additionally, this process allows water recuperation by condensation of the steam produced in the drying process.

References

[1] Ben Sassi, A., Ouazzani, N., Walker, G.M., Ibnsouda, S., El Mzibri, M. & Boussaid, A., Detoxification of olive mill wastewaters by Moroccan yeast isolates. *Biodegradation*, **19(3)**, pp. 337–346, 2008.

[2] Andreozzi, R., Canterino, M., Di Somma, I., Lo Giudice, R., Marotta, R., Pinto, G. & Pollio, A., Effect of combined physico-chemical processes on the phytotoxicity of olive mill wastewaters. *Water Research*, **42(6-7)**, pp. 1684–1692, 2008.

[3] Morillo, J. A., Aguilera, M., Antízar-Ladislao, B., Fuentes, S., Ramos-Cormenzana, A., Russell, N.J. & Monteoliva-Sánchez, M., Molecular microbial and chemical investigation of the bioremediation of two-phase olive mill waste using laboratory-scale bioreactor. *Applied Microbiology and Biotechnology*, **79(2)**, pp. 309–31, 2008.

[4] Gómez, A., Zubizarreta, J., Rodrigues, M., Dopazo, C. & Fueyo, N., An estimation of the energy potential of agro-industrial residues in Spain. *Resources, Conservation and Recycling*, **54(11)**, pp. 972–984, 2010.

[5] Russo, C., Cappelleti, G.M. & Nicoletti, G.M., LCA of energy recovery of the solid waste of the olive oil industries. *6th International Conference on LCA in the Agri-Food Sector*, Zurich, 2008.

[6] Roig, A., Cayuela, M.L. & Sánchez-Monedero, M.A., An overview on olive mill wastes and their valorization methods. *Waste Management*, **26**, pp. 960–969, 2006.

[7] AIDA – Associação Industrial do Distrito de Aveiro, Estudo Multi-Sectorial na Área do Ambiente, 61p, 2000.

[8] Gil, L., Cork powder waste: an overview. *Biomass and Energy*, **13**, pp. 59–61, 1997.

Bioenergy for regions: alternative cropping systems and optimisation of local heat supply

C. Konrad[1], B. Mast[2], S. Graeff-Hönninger[2], W. Claupein[2], R. Bolduan[1], J. Skok[1], J. Strittmatter[1], M. Brulé[1] & G. Göttlicher[3]
[1]*European Institute for Energy Research EIFER, Germany*
[2]*Universität Hohenheim, Germany*
[3]*Energie Baden-Württemberg AG (EnBW), Germany*

Abstract

In the framework of a research project for the energy supplier "Energie Baden-Württemberg AG" (EnBW) in Germany, the aim of the study is to evaluate the potentials of alternative substrates and their viability for biogas conversion based on current production regimes in the county of Biberach in the South-West of Germany. The project includes 5-year field tests of optimized cropping systems leading to higher biodiversity and sustainability while ensuring a constant biomass supply for biogas production. Furthermore, precise calculations and estimations of the heat demand of rural areas have been carried out on an object-based level (residential and tertiary/industry) using a geographic information system. On the basis of existing biogas plants, techno-economical analysis of heat and micro gas networks have been performed. Sustainability is mainly emphasized on the basis of the aspect of environmental influence on cropping systems (biodiversity, soil erosion, ground and surface water pollution). Biogas yield data at laboratory scale are used to evaluate the economy of alternative cropping systems with regard to energy production as compared to the reference (maize monoculture) in the whole chain ranging from field cultivation to energy use. The practical feasibility and the environmental effects are reviewed in comprehensive and multi-field tests and field trials.
Keywords: biomass potential, yield model, GIS, biogas, substrate, biodiversity, heat demand, building stock, heat sinks, small district heating, micro gas grid.

1 Introduction

Biomass production for the conversion into biogas has become very popular since the establishment of the renewable energy law (EEG) in 2001. Currently there are already 5700 existing biogas plants in Germany. Most biogas plants run on maize as a main substrate, cultivated in monoculture over large areas. The one-sided specialisation on a specific crop goes along with several environmental problems such as the loss of biodiversity, soil erosion and the pollution of ground and soil water.

As the EEG generously supports the generation of renewable electricity, many biogas units are operated without using enough of the heat produced despite the combined heat and power (CHP) bonus. This research project focuses on finding alternatives cropping systems and alternative use of heat for a more sustainable development of current and future biogas installations.

2 Estimation of the biomass potential

According to the official statistics, about 10% of the agricultural land was used for the cultivation of energy crops in the county of Biberach in 2007 [5]. Meanwhile, this value has significantly increased to 15%. From an ecological point of view, the problem is the increasing land use and thereby the competition with food production and animal feed production. The continuous and intensive use of agricultural areas with all the environmental consequences could lead to future deterioration of biodiversity. Actually the area used for the growth of energy crops is further increasing, focusing particularly on maize production. In relation to the entire cultivated area, the proportion of maize in 2009 reached about 30% in the county of Biberach. In several communes, even over 50% of cultivated land is used for the growth of silage maize [4]. The tendency is that this maize is the major substrate for biogas production. The rate of grassland usage for biogas production, which reaches only 3%, is still quite low and surely remains an interesting potential which is not sufficiently used. The introduction of the public subvention "manure bonus" led to an extra created incentive in 2009 in the amendment of the renewable energy law (EEG). The use of slurry and manure builds up a further potential, which amounts to about 90,000 livestock units which can be partly used for the biogas production.

The estimation of the biomass potential is based on publications of the ministry of agriculture and on the official statistics of Baden-Württemberg [4, 5]. Moreover high-resolution remote sensing data is available to the project [3]. This data allows an accurate spatial differentiation of arable- and grassland units. The calculation of biogas and methane yields is based on empirical values of on-farm biogas plants [2]. In addition to the common energy crops like silage maize, grass silage and whole crop silage (WCS), the extended use of manure as a function of livestock units is considered (fig. 1).

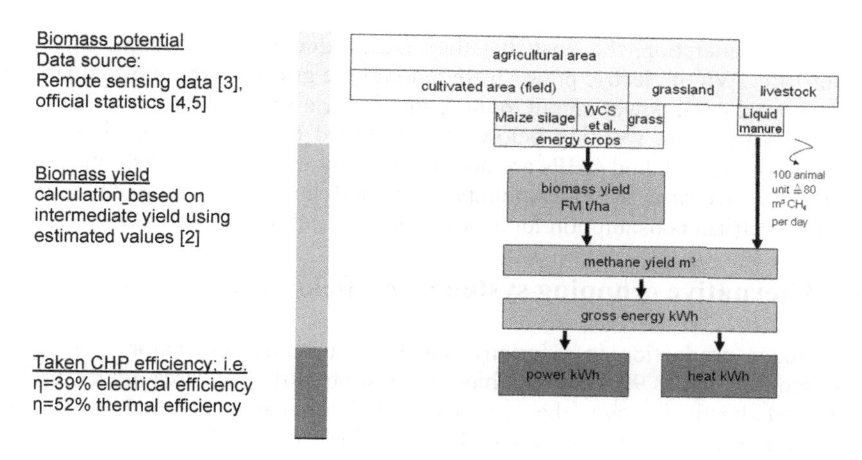

Figure 1: Schematic diagram of the estimation methodology of the biomass potential in the county of Biberach.

2.1 Status quo of biogas production

The characterisation of the installed biogas plants is based on information from the administration and the public database of EnBW [1]. Within this data the location and actual amount of electricity generated from biogas are documented. In the year 2009 74 biogas units with a rated power output of over 20 MW_{el} are already installed in Biberach. About two-thirds of the biogas plants are equipped with a combined heat and power system CHP. This means that besides the

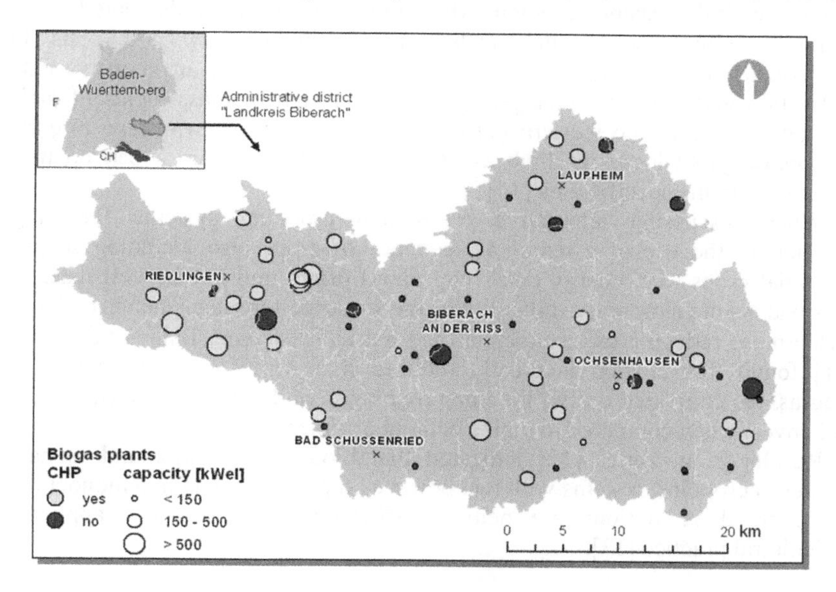

Figure 2: Location and capacity of biogas plants.

electricity generation, the heat that they are producing is also used. In 2009, about 140 GWh of electric power were fed into the grid whereby the comparison of registered rated power output to the actual electric production shows that most of the facilities are working below their installed capacities. This shows that fermenters and installed CHPs are not always fully optimised to each other. The equivalent electrical supply amounts to 50,000 households with an average annual electrical consumption for a two-person household of 2800 kWh.

3 Alternative cropping systems and field experiments

For biogas production in Germany maize is currently the major crop used, representing up to 90% of the biogas substrate [6]. The area under maize increased drastically over the last years; in 2009 maize intended for biogas production was cultivated on about 380,000 ha in Germany [7].

From an economical perspective, maize is a favorable crop on many sites, providing high energy yields per area and furthermore being an established crop with wide experience in cultivation and ensilage [8].

On the other hand, maize cropping systems are highly criticized. The strong focus of the agricultural practice on maize as the main biogas substrate entails also negative ecological impacts like soil erosion, soil organic matter reduction, nitrate leaching and the loss of biodiversity and habitats [9]. The Sachverständigenrat für Umweltfragen (German Advisory Council on the Environment) [6] estimated in its special report that maize is a crop with a high risk for nutrient leaching, soil erosion and biodiversity. Besides the effect on biodiversity, also the typical local agricultural biodiversity and landscape appearance is decreasing in regions where biogas plants are implemented.

Furthermore, diseases like *Helminthosporium turcicum* and pests like *Diabrotica virgifera virgifera* are increasing as a result of an intensive maize cultivation and thus are endangering stable and high biomass yields of maize. The occurrence of the western corn rootworm (*Diabrotica virgifera virgifera*) may cause yield losses as high as 30% and often results in a ban on maize cultivation in the specific area [10].

When discussing alternative crops and cropping systems for biogas production, the use of a temporal sequence of crops also including strips of perennial crops (e.g. energy dock, cup plant) offers multiple harvest dates and thus widens the temporal availability of the substrate for biogas plants and might furthermore reduce erosion potential and nutrient leaching. Laloy and Bielders [11] found that erosion is greatly reduced if a winter cover crop (rye and ryegrass) is cultivated during the intercropping period before maize (maximized soil cover) when compared to maize without a winter cover crop.

Results from Vetter [12] indicated that biodiversity (flora and fauna) is greater in cropping systems with two or three crop species than in a monoculture. However, these options are neither sufficiently perceived nor applied in agricultural practice [13].

3.1 Implementation of alternative cropping systems in the study region

The objective of this study lies in developing, testing and monitoring sustainable cropping systems offering a long-term alternative to maize as a biogas substrate in the region. The experimental field site of 5.6 ha is implemented on a farm located in the county of Biberach. The developed cropping system will be cultivated in this field with assistance from the farmer and different environmental relevant parameters will be collected and measured over four years. In 2010 the existing maize monoculture was monitored to determine the status quo situation, as a reference for conventional biogas crop cultivation. In the following four years, this field will be used for the cultivation of the intended alternative cropping system.

The following parameters are identified as relevant and measured:

- Fresh and dry matter yield of the different crops;
- Biogas output of the different crops;
- Erosion potential;
- Nitrate leaching into the groundwater;
- Biodiversity (arable weed species, along transects which are distributed across the field);
- Biodiversity (ground beetles, with pit fall traps, which are distributed across the field).

3.2 Developed cropping system

The developed cropping system consists of a strip-wise cultivation of a perennial crop and annual crops in a crop rotation. Fig. 3 shows the developed cropping system in comparison to a maize monoculture (M) without permanent soil coverage. For the perennial crop (pC) energy dock also known as *Rumex Schavnat* or *Rumex OK2* will be used. It is a frost-resistant crossbreed from the Ukraine with a cultivation period of 15–20 years and a potentially high biomass yield. Its biomass can be harvested twice a year and ensiled [14].

The implemented crop rotation (CR) includes the crop species sunflower (1 yr + 4 yr) – winter triticale (2 yr) – clover grass (2 yr + 3 yr) – amaranth (3 yr) – forage rye (4 yr), and it is set up completely without maize. As a further advantage, the developed cropping system allows a permanent soil cover with all the above-mentioned positive aspects. From the second year onwards energy dock could provide considerable higher biomass yields than other perennial crops. Furthermore the production costs are lower (seed, fertilizer, etc.) after the establishment phase in the first year. Strips of CR and pC will be of 24 m width and fit into the working width of the farmer's equipment.

The given combination and parallel cultivation of winter crops, summer crops and clover grass are expected to entail the above mentioned advantages and distribute the substrate supply over the year. This has a further positive effect of reducing silage storage capacity requirements.

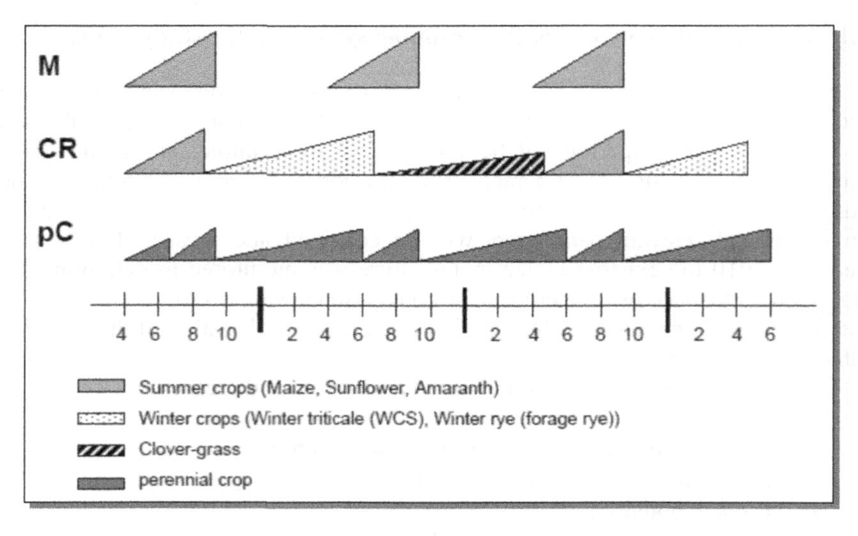

Figure 3: The cropping scheme (one run of the crop rotation) for the monoculture maize (M), the alternative crop rotation (CR) and the perennial crop (pC) (the axis of abscissa are the month of the years).

4 Model for the biogas production from energy crop mixtures

4.1 Model objectives

Energy crops of different varieties are amenable to anaerobic digestion. The crops which are most suitable for biogas production may be characterized with the following parameters:

1. High biomass yield per hectare (tons of dry mass = DM harvested per ha);
2. Low cultivation costs;
3. High digestibility (i.e. low fibre and lignin contents);
4. Appropriate C:N ratio (i.e. ranging between 20 and 30) [15].

Because of its ability to fulfill these requirements, maize is currently the most wanted energy crop for anaerobic digestion [15]. However, the environmental drawbacks of maize motivate the search for alternative energy crops.

A model, based on the identified alternative cropping systems (Section 3) was developed in order to perform an assessment of energy crops for biogas production while comparing different crop varieties. For this purpose, data were collected on methane yields of energy crops (laboratory batch test) as well as on the biomass yield per hectare. Using the composition and weight ratios of energy crops in the digester feedstock as input parameters, the feedstock mixture composition can be calculated and compared to empirically set optimal criteria (C:N ratio, fibre and lignin content) for digester operation. Moreover, the methane yield per hectare of field can be determined.

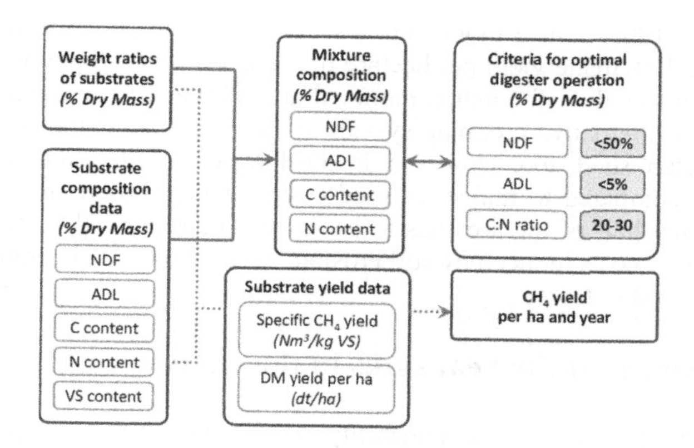

Figure 4: Input parameters (on the left) and output parameters of the model (on the right). NDF: neutral detergent fiber, ADL: Acid Detergent Lignin.

4.2 Exemplary energy crop mixture

Within this model a conventional cropping system based on maize monoculture was compared to an alternative and more sustainable cultivation system. The most methane yielding mixture was selected from the previously proposed cropping system (fig. 3). The high-yielding sustainable crop mixture was composed of 40% Triticale, 40% Cup plant and 20% Amaranth (all shares on a dry matter basis). Fig. 5 shows the dry matter yield (on the left) and the methane yield (on the right). For both parameters maize, the alternative cropping system as well as the single crops is pictured.

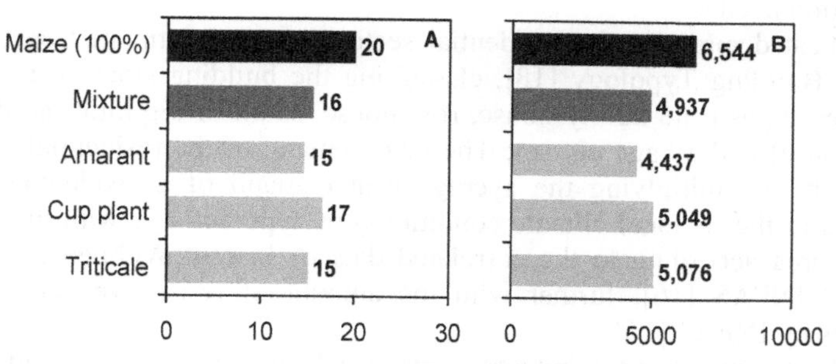

Figure 5: A. Dry matter yield (t/ha) B. CH_4 yield (m^3/ha).

In this case, the empirical criteria set for low fiber content (Neutral Detergent Fiber = NDF<50%), low lignin content (Acid Detergent Lignin = ADL<5%) and balanced nutrient content (C:N ratio in the range 20–30) were fulfilled. One may deduce that full-scale biogas units may be successfully operated with such manure-free energy crops mixture. The calculated methane yield of the mixture was almost 5000 m^3_{CH4}/ha/year. In comparison, the calculated methane yield per hectare for a 100% maize monoculture reached 6500 $m^3{}_{CH4}$/ha/year. Therefore the theoretical methane yield per hectare of the alternative cropping system was about 20% lower than for maize monoculture. The cultivation costs might be higher for the alternative cropping system in the first year, as the perennial crop has to be established, too. However, in the long run, cultivation costs will be lower when compared to maize, as seed costs for the chosen annual crops are lower and only half of the area has to be sown. The rough calculation does not include indirect costs induced by soil erosion, loss of biodiversity and the impact of pesticides and fertilizers.

5 Assessment of the heat sinks in the study area

Besides the increase of sustainability with alternative cropping system (Section 3), prudent heat usage of existing and future biogas plants shall be investigated and alternatives be prospected. The essential input is the estimation of the heat demand in communes in the county of Biberach. Further, the identification of main heat sinks with potential for different grid-bound heat supply concepts, based on biogas, was focused.

5.1 Methodology of the heat demand assessment

For the heat demand estimation, existing building stock data, statistic data for regional economy and energy consumption by sector for the district Biberach, were analyzed. For the residential and the tertiary/industry sector geo-referenced INFAS data for building stock [16] (age, class and number of households) was used. Additionally the database of the regional economy (FIS), containing companies' addresses and numbers of their employees, arranged by the regional Chamber of Industry and Commerce [17] was included, to complete the information needed.

The heat demand of the residential sector has been estimated based on the German Building Typology [18], classifying the building stock in a matrix of 5 building types (one-family house, row house, small or big multifamily house, tower house) and 10 age classes. The heat demand of the residential sector has been built by multiplying the average heat demand of a residential building (adjusted to the regional climate conditions) by type and age with the building' average area according to the correlated data of Institut Wohnen und Umwelt [18] and INFAS [16]; further with the amount of respective geo-referenced buildings by INFAS [16].

The heat demand of the industry sector has been received by consideration of the statistic data on the energy consumption according the national branches taxonomy [19]. The specific energy consumption in branches on state level (without electric power) has been accounted referring to the number of employees per branch [20]. It amounts e.g. 22 MWh/employee in machines manufacturing, 8 MWh/employee in furniture construction or 379 MWh/ employee in paper production). This specific value has been multiplied by the average employee number per enterprise according the FIS data base [17] and geo-referenced by addresses.

Major regional heat sinks in the tertiary sector (trade, commerce and services) have been partially considered, i.e. kindergarten, schools, swimming pools have been geo-referenced according the INFAS data [16] and their heat demand has been estimated with help of average values (surface and heat demand) according to current literature [21–23].

5.2 Heat demand in the region

The built-up areas with the heat demand of buildings, aggregated in a 100 m raster, have been calculated and localized in a geographic information system (GIS). The values of the heat demand raster varied here between zero (by raster points without any buildings) and max. 150.000 MWh/a – doing so, the local repartition of heat sinks could be obtained (Figure 6).

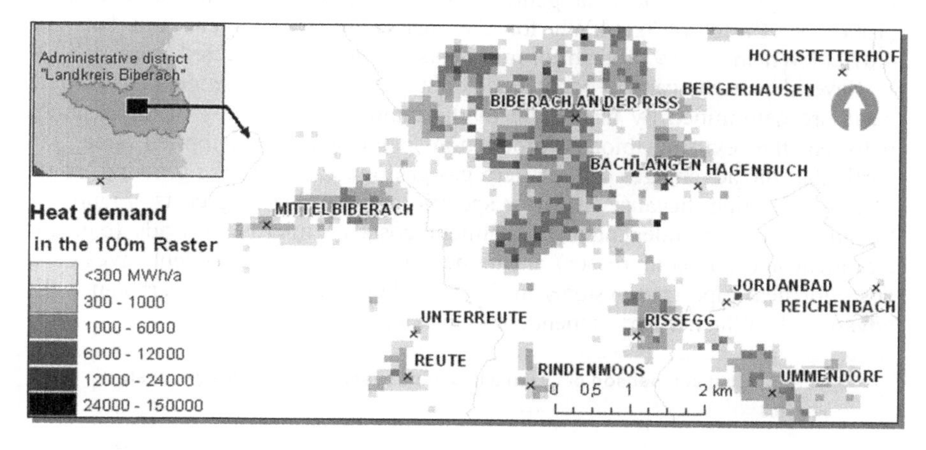

Figure 6: Detail map of the heat demand in raster of 100 m in communes of the Biberach district.

Additionally, a ranking of communities of the county of Biberach, ordered by their annual heat demand, has been worked out: in 73% of all 45 considered communities, the estimated heat demand was below 100 GWh/a. Eight communities (17%) showed values between 100 GWh/a and 200 GWh/a. The highest heat demand of 1,155 GWh/a has been accounted for the county centre Biberach, followed by the estimated heat demand for Laupheim of 470 GWh/a.

In the following the heat demand density of built-up areas within the examined communities was identified as the value of the aggregated heat demand raster points in relation to the surface of the built-up areas. It shows values between 4 MWh/ha*a and 5.680 MWh/ha*a. For almost two third of the built-up areas the heat demand density was estimated below 300 MWh/ha*a, over one-fourth showed values of between 300 and 700 MWh/ha*a.

6 Techno-economic analysis of alternative heat supply systems for model communities in the Biberach district

For the purpose of investigating heat networks (district heating supplied by biogas) with micro gas grids (autonomous biogas grid from the biogas plant to the heat demand site) a techno-economic calculation model has been established. Three rural areas have been selected, according the following criteria: heat demand density bigger than 300 MWh/ha*a (see Section 5), distance to biogas plants under 1,000 m (as pre-conditions for all model areas). Further, the availability of a gas network and a location in a neighborhood of an industry zone was considered. The most relevant parameters influencing heat generation costs could be identified.

In respect of the heat demand conditions in the considered model areas (different access rate of buildings), the optimal solution for each study case could be shown. The specific heat generation costs varied between 4.47 € ct/kWh (100ct = 1€) to 14.52ct/kWh for the heat pipe grid system and between 4.87 ct/kWh and 14.49 ct/kWh for micro gas grid in the considered areas.

Several patterns can be concluded. First, the network-dependent investment costs are determined by the length of the main pipe (in terms of the distance between the existing biogas production plant and the supplied area). In particular, the advantages of the micro gas grid increase with the growing length of the main pipe, because of lower specific civil engineering costs (Figure 7). Secondly, the net-independent investment costs of micro gas grids (e.g. gas treatment and gas compressor) are higher than the net-independent investment costs for heat pipe grid systems in the considered cases - they overbalance the advantages of the network-dependent investment costs.

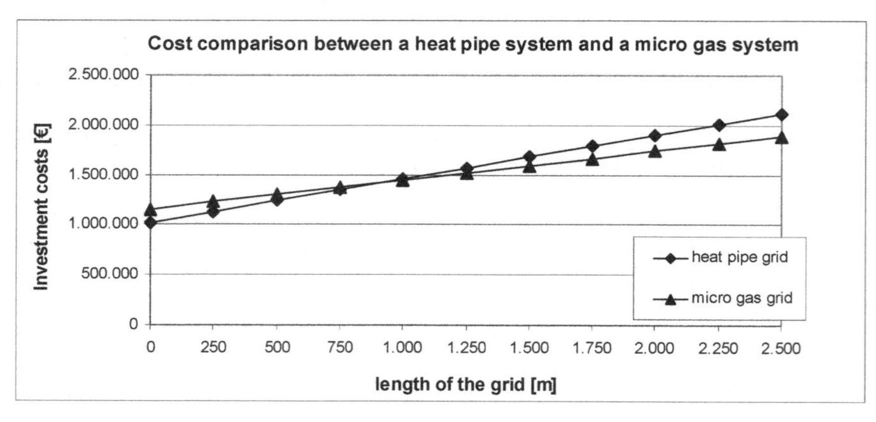

Figure 7: Comparison of investment costs between a micro gas grid and a heat pipe grid in relation to their length.

In most considered cases, the biogas potential can cover only partially the local heat demand, the remaining heat demand has to be supplied by other fuels and technologies (e.g. wood or gas combustion in a peak load boiler).

7 Conclusion

The project shows a high potential in alternative cropping systems and in the optimization of the local heat supply. In the following years, new crop mixtures will be tested on field and their competitiveness with the reference system maize monoculture will be economically calculated. The impacts on the environment will be monitored in a long-term study. Expected advantages concern erosion reduction, fertilizer demand and nutrient leaching mitigation as well as a decreasing use of pesticides or reduced pest infestation. Furthermore, biodiversity may be enhanced and storage capacity requirements for ensiling harvested crops may be reduced.

The heat demand analysis provides a precise picture of the potential of existing heat sinks. The techno-economic analysis of heat and gas networks shows the viability of the systems in relation to the distance, the demand site and the local biomass potential.

References

[1] EnBW: EEG-Anlagendaten in der Regelzone der EnBW Transportnetze AG, http://www.enbw.com

[2] Faustzahlen Biogas, Kuratorium für Technik und Bauwesen in der Landwirtschaft e. V. (KTBL), 2008.

[3] Mapping crop distribution in administrative districts of southwest Germany using multi-sensor remote sensing data, Department of Remote Sensing, Institute of Geography, University of Würzburg, 2010. http://dx.doi.org /10.1117/12.865113

[4] MLR, Ministerium für Ländlichen Raum, Ernährung und Verbraucherschutz, 2010.

[5] Struktur- und Regionaldatenbank, Statistisches Landesamt Baden-Württemberg, 2010. http://www.statistik.baden-wuerttemberg.de/

[6] Sachverständigenrat für Umweltfragen, *Umweltschutz durch Biomasse – Sondergutachten, Juli 2007,* Erich Schmidt Verlag & Co KG: Berlin, pp. 43–56, 2007.

[7] Deutsches Maiskomitee, Maisanbaufläche Deutschland in ha, 2008 und 2009 (vorläufig) nach Bundesländern und Nutzungsrichtung in ha. http://www.maiskomitee.de/web/upload/pdf/produktion/Maisanbauflaeche_D_08-09.pdf

[8] Gömann, H., Kreins, P. & Breuer, T., Deutschland – Energie-Corn-Belt Europas? *Agrarwirtschaft*, **56(5/6)**, pp. 263–271, 2007.

[9] Miehe, A. K., Herrmann, A. & Taube, F., Biogaserzeugung aus landwirtschaftlichen Rohstoffen – Monitoring des Substratanbaus und Gärrestverwertung in Schleswig-Holstein. *Mitteilung der AG Grünland und Futterbau Band 9, Referate und Poster der 52. Jahrestagung der AG Futterbau und Grünland der Gesellschaft für Pflanzenbauwissenschaften,* eds. Thomet, P., Menzi, H. & Isselstein, J., AGRIDEA, pp. 313–316, 2008.

[10] Schwabe, K., Kunert, A., Heimbach, U., Zellner, M., Baufeld, P. & Grabenweger, G., Der Westliche Maiswurzelbohrer (*Diabrotica virgifera virgifera* LeConte) – eine Gefahr für den europäischen Maisanbau. *Journal für Kulturpflanzen*, **62(8)**, pp. 277–286, 2010.

[11] Laloy, E. & Bielders, C. L., Effect of intercropping period management on runoff and erosion in a maize cropping system. *Journal of Environmental Quality*, **39(3)**, pp. 1001–1008, 2010.

[12] Vetter, A., *Standortangepasste Anbausysteme für Energiepflanzen – 3. Auflage*. Fachargentur Nachwachsende Rohstoffe e.V: Gülzow, pp. 79–98, 2010.

[13] Willms, M., Glemnitz, M. & Hufnagel, J., FNR-Projekt „Entwicklung und Vergleich von optimierten Anbausystemen für die landwirtschaftliche Produktion von Energiepflanzen unter den verschiedenen Standortbedinungen Deutschlands (EVA) "Teilprojekt II "Ökologische Folgewirkungen des Energiepflanzenanbaus". http://www.tll.de/vbp/eva1 /zalf_tp2.pdf

[14] Vetter, A., Heiermann, M. &Towes, T., *Anbausysteme für Energiepflanzen – optimierte Fruchtfolgen + effiziente Lösungen,* DLG-Verlags-GmbH: Frankfurt am Main, pp. 81, 2009.

[15] Weiland, P., Biogas production: current state and perspectives. *Applied Microbiology and Biotechnology*, **85**, pp. 849–860, 2010.

[16] INFAS; http://www.infas-geodaten.de/

[17] Data Base of Regional Economy – Firmeninformationssystem (FIS) Firmendatenbank des Baden-Württembergischen Industrie- und Handelskammertages (IHK), http://www.bw-firmen.ihk.de/sites /fitbw /search/detailSearch.aspx

[18] Institut Wohnen und Umwelt (IWU) Darmstadt; Deutsche Gebäudetypologie; Systematik und Datensätze; Dezember 2003.

[19] Energieverbrauch des Verarbeitenden Gewerbes, Bergbau und Gewinnung von Steinen und Erden in Baden-Württemberg 2008; Statistisches Landesamt Baden-Württemberg, 2010.

[20] Verarbeitende Gewerbe, Bergbau und Gewinnung von Steinen und Erden in Baden-Württemberg 2008; Statistisches Landesamt Baden-Württemberg, 2009.

[21] Schlomann, B., Gruber, E., *et al.*; Energieverbrauch des Sektors Gewerbe, Handel, Dienstleistungen (GHD) für die Jahre 2004 bis 2006 Projektnummer 45/05 Abschlussbericht an das Bundesministerium für Wirtschaft und Technologie (BMWi) und an das Bundesministerium für Umwelt, Naturschutz und Reaktorsicherheit (BMU); Karlsruhe, München, Nürnberg; 2009.

[22] Bundesministerium für Verkehr, Bau und Stadtentwicklung (BMVBS); Benchmarks für die Energieeffizienz von Nichtwohngebäuden, Vergleichswerte für Energieausweise; Berlin; 2009.

[23] VDI-Richtlinie 3807 Blatt 1 Energieverbrauchskennwerte für Gebäude; Tabelle 4, 1998.

The behavior of suspended particulate matter emitted from the combustion of agricultural residue biomass under different temperatures

Q. Wang, S. Itoh, K. Itoh, P. Apaer, Q. Chen, D. Niida, N. Mitsumura, S. Animesh, K. Sekiguchi & T. Endo
Graduate School of Science and Engineering, Saitama University, Japan

Abstract

There are large quantities of waste rice husk and straw estimated at around 3.9 million tons as biomass waste every year in Japan. Air pollutants emitted from exhaust gases of rice husk incineration lead to environmental damage, not only because of the influence on global environment and climate, when released into the atmosphere, but also on human health due to local air pollution. Therefore, it is necessary to effectively utilize waste rice husk and straw to reduce air pollutants. In recent years, there has been an increasing demand on the utilization of unused biomass instead of fossil oil fuel in combustors for farming-greenhouses' heating during the winter season. The increasing demand will increase the running costs. In general, since these combustors are small in size, there is a lack of regulations or laws (e.g. The Air Pollution Control Act and The Waste Disposal and Public Cleaning Law) in operation for their air pollution control. So far, small size combustors are characterized by their simplicity of structure and low costs. However, they emit visible black carbon (elemental carbon) due to their poor combustion performance. In this study, the possibility of the substitution of fossil fuel by waste rice husk and rice straw is investigated in laboratory model combustion experiments. The emission behavior of harmful air pollutants emitted from rice husk and straw combustion is evaluated by measuring carbonaceous and ionic composition of suspended particulate matter in the exhaust gases. From the analytical results, particulate mass concentrations were found reduced substantially at high temperature combustion. From the results of this study, it can be suggested that stable combustion performance

under suitable conditions is needed in order to control less air pollutants emitted from biomass fuel although small size combustors are still not regulated.
Keywords: rice husk and rice straw, small size combustors, combustion conditions, carbonaceous, ionic and metallic composition, $PM_{2.5}$.

1 Introduction

Global warming has become increasingly evident in the global climate. Combustion of fossil fuel is generally admitted as the main cause for global warming. However, the use of fossil fuel is expected to increase in the future because of economic development and growth of population in developing countries [1], hence, the only solution is zero-emission technology, that is, to reduce all possible emissions produced by human activities to zero [2]. In order to achieve zero-emissions, it is important to apply a technology to utilize all unused biomass [3; 4].

Currently in Japan agriculture and forestry produce biomass residues from where a very small amount is used, and unused biomass is mostly being incinerated for disposal, due to its high cost of collection, transport, and storage and also the needs of energy that it implies. Moreover, urgent countermeasures are required to reduce the air pollution from illegal waste biomass incineration. It is estimated that only in Japan around 3.9 million tons of waste rice husk and rice straw, which is the most common agricultural residue in the country, are wasted every year. Additionally, since rice is the staple food and regular part of the diet for almost half of the world population, an effective utilization of waste rice husk and straw as biomass fuel would be an important countermeasure to global warming. In recent years there is an increasing demand on the utilization of unused biomass instead of usual fossil oil fuel combustors for farming-greenhouses heating during the winter season. This increase in the demand will make prices to increase. In general, these combustors are small in size [5] therefore existing regulations do not apply (e.g. the air pollution control act and the waste disposal and public cleaning law). So far, small size combustors are characterized by simplicity in their structure and low costs, however, visible black smoke and pollutants are emitted due to the poor combustion performance and the lack of regulations [6, 7]. In this study, the potential use of waste rice husk as biomass fuel is investigated based on laboratory model combustion experiments. Firstly, the chemical composition of agricultural waste rice husk and straw were analyzed thus investigating its combustion characteristics. Then, the air pollutants emitted from waste rice husk and straw combustion were measured by sampling suspended particulate matter and gases in the exhaust under the different combustion conditions. The possibility of reduction of these harmful substances in $PM_{2.5}$ and the exhaust gases were also investigated.

2 Experimental methods

2.1 Composition analysis of the rice husk and straw

In this study, rice husk and straw composition (Nigata Prefecture of Japan) were analyzed. The proximate and ultimate analyses of their samples were carried out according to the Japanese industrial standard (JIS) method of JIS-M8812 and JIS-M8813.

2.2 Evaluation method of the combustion characteristics of the rice husk and straw

Combustion characteristics of rice husk and straw were analyzed by the thermogravimetric/differential thermal analysis (TG/DTA, Model DTG-60, Shimadzu Co. Ltd., Japan), and under the following conditions: samples were prepared below 250 µm by several sieves. About 4.0 mg of samples were heated at a rate of 5°C min^{-1} starting from room temperature until 900°C. A gas flow rate of 250 mL min^{-1} was used; clean gas was used as the carrier gas for combustion.

2.3 Evaluation of suspended particulate matter (SPM) in exhaust gases

2.3.1 Air sampling method for method for exhaust gases collection

Biomass burning is an important source of primary fine particles in the atmosphere, which can influence the regional air pollution and human health. Recently, fine particles below 2.5 µm in aerodynamic diameter (e.g. $PM_{2.5}$) either emitted from biomass burning or generated by photochemical reactions, are of great concern because of their effect on health and environment in Japan. For example, coarse particles of suspended particulate matter are mainly having particle sizes larger than 2 µm and are unable to instruction in to entering the respiratory tract by the nose, throat, and pharynges. Therefore, in this study, the $PM_{2.5}$ from combustion of waste rise husk and straw is evaluated. The collection devices of exhaust gases are shown in fig. 1. In order to evaluate the $PM_{2.5}$ emissions for the combustor, exhaust $PM_{2.5}$ were collected on quartz-fiber filters (35 mm φ, Pallflex Products Corp, 2500QAT-UP) and Teflon filters (35 mm φ, Pallflex Products Corp, T8711A) using two air samplers which are called the $PM_{2.5}$ personal sampler (Model NWPS-35HS, Sibata Scientific Technology Co. Ltd., Japan). The quartz-fiber filter was used for carbonaceous and ionic composition analysis, and the Teflon filters were used for metal composition analysis. Gaseous components (CO, CO_2, NOx and O_2) were also evaluated by portable gas analyzer (Model PG-250, Horiba Co. Ltd.).

2.3.2 Evaluation of carbonaceous compositions of $PM_{2.5}$ in exhaust gases

Carbonaceous analysis was based on the IMPROVE method (Interagency Monitoring of Protected Visual Environment) by the thermo-optical carbon analyzer (thermo/optical carbon analyzer: Model 2001, Desert Research Institute) shown in table 1. In this method, a 0.503 cm^2 (8 mm diameter) punch

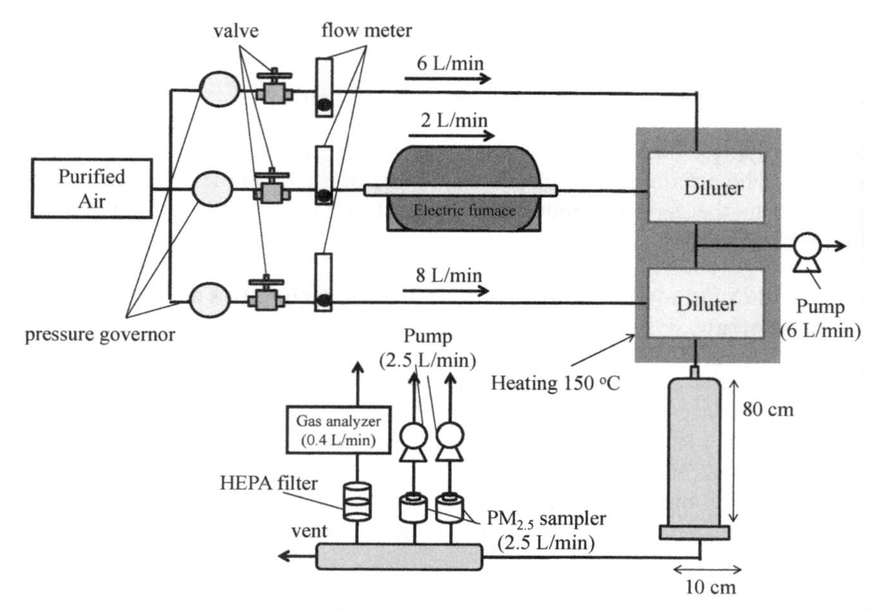

Figure 1: Air sampling setup for exhaust gases emitted from the combustor.

Table 1: Protocol of IMPROVE thermal/optical method for carbonaceous analysis.

Thermal/optical method		
Fraction	Temperature (°C)	Atmosphere
OC1	120	100% He
OC2	240	
OC3	450	
OC4	550	
EC1	550	$2\%\ O_2 + 98\%$ He
EC2	700	
EC3	800	

aliquot of a sample quartz filter was heated at 120°C (OC1), 250°C (OC2), 450°C (OC3), and 550°C (OC4) in a helium atmosphere, and then at 550°C (EC1), 700°C (EC2), and 800°C (EC3) in an oxidizing atmosphere of 2% oxygen and 98% helium. The analysis was repeated two or three times for each sample for better accuracy. $PM_{2.5}$ samples were collected at the flow rate of 2.5 L min^{-1} for the combustion out on each sampling with the $PM_{2.5}$ personal sampler.

2.3.3 Evaluation of ionic composition of suspended particulate matter in exhaust gases

One half of the 35 mm φ quartz-filter was ultrasonically extracted with 5 mL ultrapure water (18.2MΩ milli-Q ultrapure water) for 20 minutes, in order to carry the ionic composition analysis. The concentrations of the following cations were measured: Ca^{2+}, K^{+}, NH^{4+}, and Na^{+} and the following anions: SO_4^{2-}, NO^{3-}, and Cl^{-} anions and cations were analyzed in two different ion chromatographs (IC, Model DX-100, Dionex Co. Ltd., Japan).

3 Results and discussion

3.1 Measurements in the composition of waste rice husk and straw

The bulk composition of biomass in terms of carbon, hydrogen, and oxygen (CHO) did not differ much among different biomass sources. Typical dry weight percentages for C, H, and O were 30% to 60%, 5% to 6%, and 30% to 45% respectively [8]. Table 2 shows the composition analysis of waste rice husk and straw. As the results from the proximate analysis of rice husk indicated, ash contents were high in waste rice husk, while the carbon contents were low from the ultimate analysis of waste rice husk and straw. This means that waste rice husk heating value is lower than that of waste rice straw. However, waste rice husk and straw is much lower when compared to fossil fuel. Therefore, it is necessary to find the suitable combustion conditions for effective utilization as biomass fuel of waste rice husk and straw.

Table 2: Rice husk and straw composition analysis.

	Proximate analysis (wt%)				Ultimate analysis (wt%)			
Sample	M	VM	Ash	FC	C	H	N	O
Rice husk	5.4	62.5	17.5	14.6	45.11	5.87	0.52	30.99
Rice straw	5.3	69.2	9.2	16.3	39.19	5.26	0.51	45.89

M: Moisture, VM: Volatile matter and FC: Fixed carbon.

3.2 Differential combustion characteristics of the rice husk types

A similar trend was observed for rice husk and rice straw in the case of TG/DTA (figure 2). The TG/DTA thermogram for rice husk showed two well-defined peaks at 280°C and around 400°C. These results show that waste rice husk achieves its pyrolysis at around 280°C, where the more volatile components were burned while the carbonized fraction was burned at a higher temperature, around 400°C. For this reason, the waste rice husk can only be combusted under conditions at temperatures above 500°C. In this study, combustion was carried out at temperatures between 500 and 1000°C in increments of 100°C.

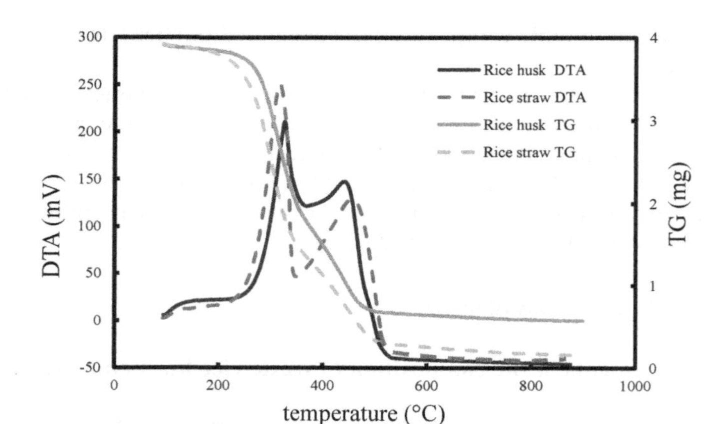

Figure 2: Pyrolysis and combustion behavior of rice husk and rice straw.

3.3 Air pollutant emitted from combustion of waste rice husk under different combustion temperatures

3.3.1 Gases components and combustion efficiencies

In this study, to simplify the combustion efficiency, account was taken of the fact that >90% of the carbon combusted in a fire was emitted in the form of CO_2 and CO, and <10% of carbon was in species such as hydrocarbons and particulate carbon. With this in mind, the modified combustion efficiency (MCE) can be defined as

$$\mathrm{MCE} = \frac{[C]_{\mathrm{CO_2}}}{[C]_{\mathrm{CO}} + [C]_{\mathrm{CO_2}}} \tag{1}$$

The combustion conditions can be categorized according to the MCE: MCE $\geqq 0.9$ indicates smoldering combustion and MCE < 0.9 indicates flaming combustion [9].

The behavior of gaseous components during waste rice husk and straw combustion under different temperatures (500–1000°C) is shown in fig. 3. It was found that all of gas concentrations showed the similar behavior in two temperature ranges 500–700°C and 800–1000°C.

Figure 4 shows the variation of the modified combustion efficiency (MCE) under all combustion conditions with the different temperatures. For example, as fig. 3 also shows, within the temperature range 700–800°C, CO and O_2 concentrations were increased under smoldering combustion conditions and NO_X and CO_2 concentrations were increased under flaming combustion conditions. These results indicate that the combustion efficiencies under flaming combustion are better than under smoldering combustion. Biomass fuel is regarded as a renewable energy source with low CO_2 emissions if produced in a sustainable manner. From this point of view, if waste rice husk and straw is used as providing a more effective fuel combustion, the emission of CO_2 is less or not contributing to global warning.

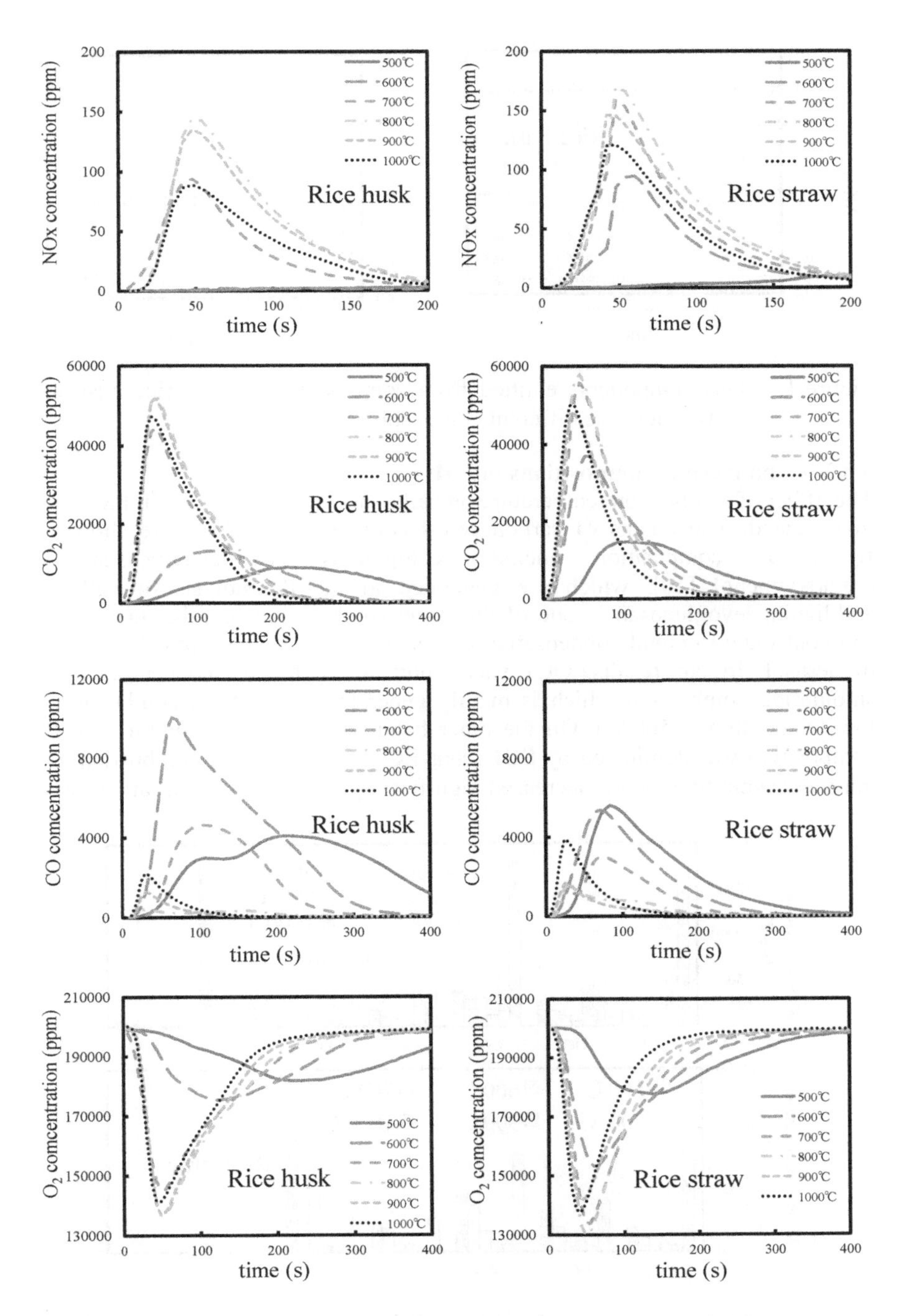

Figure 3: Gas components emitted from combustion of waste rice husk and straw under the different combustion temperatures.

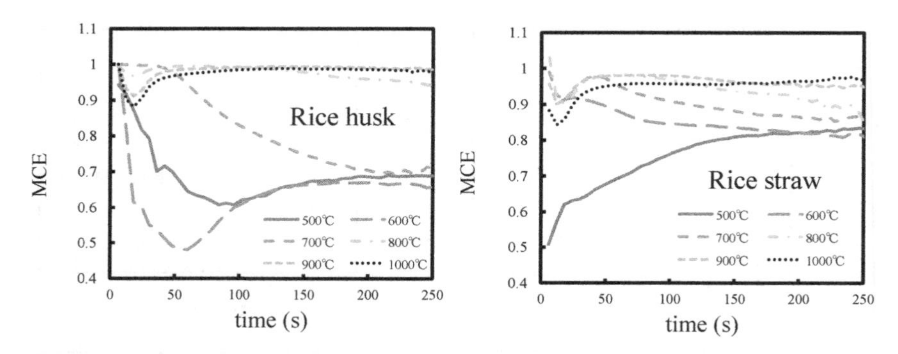

Figure 4: Gas components emitted from combustion of waste rice husk and straw under the different combustion temperatures.

3.3.2 Cabonaceous compositions in $PM_{2.5}$

The effect of combustion temperature on carbonaceous composition in $PM_{2.5}$ was investigated. The results of carbonaceous composition analysis are shown in fig. 5. OC compositions includes compounds like levoglucosan and methoxyphenol [10], which are generated in the thermolysis of cellulose and lignin; levoglucosan is one of the water-soluble organic substances and it can contribute to cloud condensation nuclei and influence the optical properties of aerosol. In our results, OC1 was found at the highest concentrations in smoldering combustion which is mainly generated by biomass combustion at low temperatures (500°C). On the other hand, in smoldering combustion, EC composition was dominated by EC1 (char-EC). Under flaming combustion, OC mass concentrations were decreased significantly, and EC concentrations were

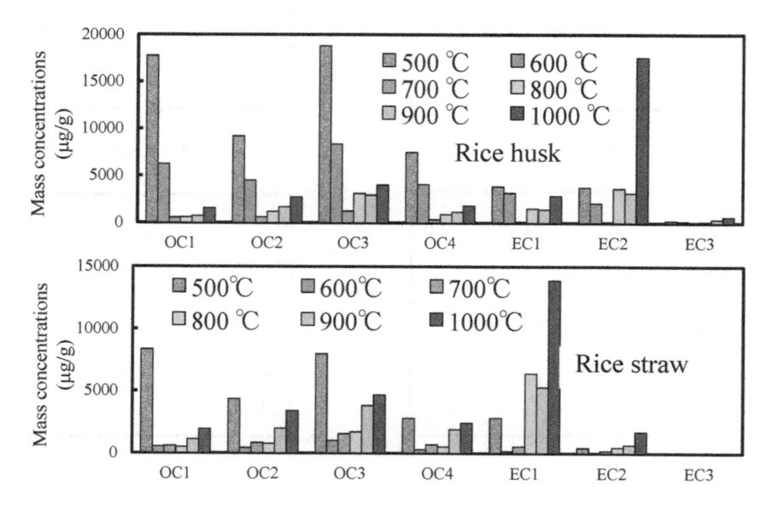

Figure 5: Carbonaceous components in $PM_{2.5}$ from combustion rice husk and straw at different combustion temperature.

dominated by EC2 (soot-EC). However, EC (EC1+EC2) were emitted by smoldering combustion. Here, EC is a mostly POC, POC is OC (OC3). Therefore, as shown in fig. 5, EC has not been nearly exhausted in smoldering combustion. From the results of total carbonaceous concentration (OC+EC), it is noted that the carbonaceous concentrations in $PM_{2.5}$ under flaming combustion were 10 times lower than under smoldering combustion.

3.3.3 Ionic compositions in $PM_{2.5}$

Ionic concentrations in $PM_{2.5}$ were almost unchanged by combustion temperature. The results are shown in fig. 6. The high concentrations of SO_4^{2-} in $PM_{2.5}$ were determined at all combustion temperatures. In general, K^+ is an important component of biomass [11], since it is used in metabolic processes; therefore, this component can be used as a marker for biomass combustion contributing to air pollution. However, in our results, K^+ concentrations were low at all combustion temperatures.

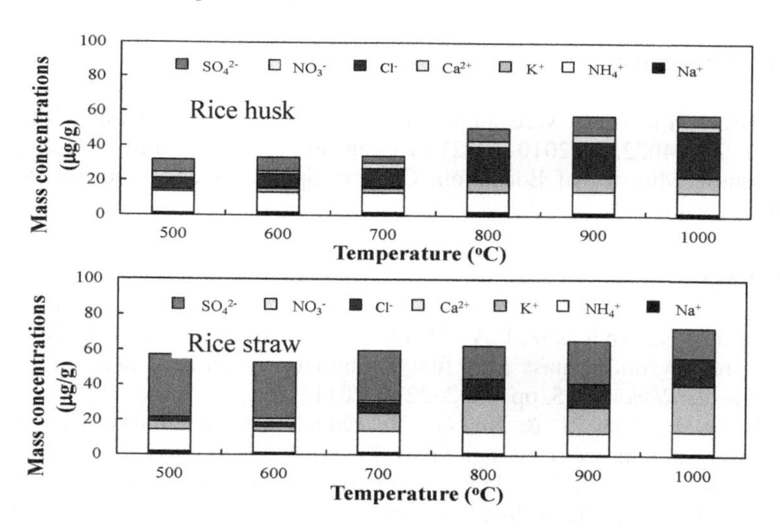

Figure 6: Ions components in $PM_{2.5}$ from combustion rice husk and straw at different combustion temperature.

As mentioned above, the behavior of harmful air pollutants emitted from rice husk and straw combustion were investigated by measuring carbonaceous and ionic composition of suspended particulate matter in $PM_{2.5}$ and the exhaust gases. It would be pleasing if this information is useful in production and application of small size combustors for waste rice husk or other biomass fuels. For further study, there is also need to analyze for polycyclic aromatic hydrocarbons (PAHs) emitted from rice husk and straw combustion in order to effectively reduce harmful air pollutants.

4 Conclusion

In this study, the possibility of waste rice husk and straw substitute fossil fuel was evaluated based on laboratory model combustion experiments. According to the combustion characteristics of rice husk and straw, it is possible to use as a biomass fuel if rice husk is combusted at the temperatures above 500°C. From analysis of gaseous compositions, it is indicated that flaming combustion (800–1000°C) proved better in its efficiency than smoldering combustion (500–700°C).

From the results of carbonaceous compositions, it was found that one tenth of carbonaceous particulate matter with smaller particle sizes may be only emitted under flaming combustion rather than under smoldering combustion. It can be suggested that air pollutants can be easily reduced if stable combustion performance under suitable conditions can be ensured especially for developing small size combustors.

Acknowledgements

Some parts of this study were supported by the special funds for Basic Research (B) (No. 22404022, FY2010–2012) of Grant-in-Aid for Scientific Research of the Japanese Ministry of Education, Culture, Sports, Science and Technology (MEXT), Japan.

References

[1] Saidur, R., Abdelaziz, E.A., Demirbas, A., Hossain, M.S. & Mekhile, S., A review on biomass as a fuel for boilers. *Renewable and Sustainable Energy Reviews,* **15**, pp. 2262–2289, 2011.

[2] Dong, L., Gao, S. & Xu, G., No reduction over biomass char in the combustion process. *Energy Fuels*, **24**, pp. 446–450, 2010.

[3] Hansen, L.A., Nielsen, H.P., Frandsen, F.J., Dam-Johansen, K., Hørlyck, S. & Karlsson, A., Influence of deposit formation on corrosion at a straw-fired boiler. *Fuel Processing Technology*, **64(1-3)**, pp. 189–209, 2000.

[4] Iliopoulou, E.F., Antonakou, E.V., Karakoulia, S.A., Vasalos, I.A., Lappas, A.A. & Triantafyllidis, K.S., Catalytic conversion of biomass pyrolysis products by mesoporous materials: Effect of steam stability and acidity of Al-MCM-41 catalysts. *Chemical Engineering Journal*, **134**, pp. 51–57, 2007.

[5] Johansson, L.S., Tullin, C., Leckner, B. & Sjovall, P., Particle emissions from biomass combustion in small combustors. *Biomass and Bioenergy*, **25**, pp. 435–446, 2003.

[6] Simoneit, B.R.T., Biomass burning – a review of organic tracers for smoke from incomplete combustion. *Applied Geochemistry*, **17(3)**, pp. 129–162, 2002.

[7] Wiinikka, H. & Gebart, R., Experimental investigation of the particle emissions from a small-scale pellets combustor. *Biomass and Bioenergy*, **27**, pp. 645–652, 2004.

[8] Khan, A.A., de Jong, W., Jansens, P.J. & Spliethoff, H., Biomass combustion in fluidized bed boilers: Potential problems and remedies. *Fuel processing technology*, **90**, pp. 21–50, 2009.

[9] Ward, D.E & Hao, W.M., Air toxic emissions from burning of biomass globally-preliminary estimates. *Proceedings of Air & Waste Management Assoc. 85th Annual Meeting & Exhibition*, 1992.

[10] Kuo, L.-J., Louchouarn, P. & Herbert, B.E., Influence of combustion conditions on yields of solvent-extractable anhydrosugars and lignin phenols in chars: Implications for characterizations of biomass combustion residues. *Chemosphere*, **85**, pp. 797–805, 2011.

[11] Jones, J.M., Darvell, L.I., Bridgeman, T.G., Pourkashanian, M. & Williams, A., An investigation of the thermal and catalytic behavior of potassium in biomass combustion. *Proceedings of the Combustion Institute*, **31**, pp. 1955–1963, 2007.

Biomass char production at low severity conditions under CO_2 and N_2 environments

G. Pilon & J.-M. Lavoie
Université de Sherbrooke, Canada

Abstract

In a perspective of biomass value addition, biomass char, a thermochemical product, long time considered as a residue, is now getting attention and may represent a vector in the sustainability of the whole biomass sector. Some char types were shown to have great potential as a solid fuel or precursor for further transformations as well as having attributes for storage and transportation. Other types showed potential as a soil carbon sequestration technique and soil amendment enhancing biomass yields. Depending on several factors, but mostly on biomass and production conditions, biomass char physico-chemical characteristics may vary tremendously. Therefore, in order to be used in accordance for specific utilizations, its characteristics must be carefully understood and controlled. In this study, chars with varying properties were produced in a custom-made lab-scale fixed bed reactor. Along these experiments, various biomass chars were produced under CO_2 and N_2 for temperatures of 300 and 500°C. Char was produced from switchgrass (*Panicum virgatum L.*) an energy crop grown in Canada. It was then characterized for ultimate and proximate analysis as well as for calorific value. In addition, specific surface was characterized by Brunauer-Emmett-Teller (BET) technique. Char organic content composition was verified by Soxhlet extractions using dichloromethane and extracts were analysed by GC-MS.
Keywords: BET, biochar, biomass, bio-oil, char, CO_2, pyrolysis, torrefaction.

1 Introduction

Biomass char refers to biochar with respect to soil amendment, and to charcoal when referring to the charred organic matter used as a source of energy [1, 2]. The latter has been shown to be a potential soil amendment that would enhance biomass production/hectare. It was also shown to have positive effects on soil health as well as being able to store carbon for reduction of CO_2 emission [3, 4]. Another application, potentially complementary to soil amendment [5], stated that biomass char also had potential to provide the required heat for pyrolysis' endothermic reaction along bio-oil production. Furthermore, it was shown to be advantageous as a feedstock for combustion or for syngas production through gasification [6, 7].

Biomass char can be obtained through different thermochemical processes operating at low oxygen content using varying degrees of severity. In chronological order of severity, torrefaction, slow and fast pyrolysis and gasification are processes under which biomass char can generally be obtained. Torrefaction can be defined as the conversion of lignocellulosic material occurring without oxygen, within a temperature range of 200 to 300°C at atmospheric pressure [7, 8]. It consists mostly of hemicellulose degradation resulting in a solid material with low moisture content, hydrophobic properties and high calorific value if compared to original biomass. Most likely, biomass char obtained under torrefaction may not meet biochar stable soil characteristics since original biomass requires sufficient temperature (≥300°C) in order to produce carbonized material with modified chemical bonds, which are less likely biodegradable [2]. Slow pyrolysis operates usually at temperatures between 300 and 500°C, at heating rates lower than 2°C/s and long residence time (min to days), conditions that were shown to enhance char production over non-condensable gases and oil. Fast pyrolysis operates at higher heating rate (up to 10 000°C/s), shorter residence time (<1 s to 30 s) and enhances biooil over char and non condensable gases yields [9, 10]. Gasification conditions operate with low-mid level of oxygen, yielding generally up to about 5–10% biomass char. Despite their lower char production yield compared with slow pyrolysis, the potential of fast pyrolysis and gasification char were recently considered as having a soil amendment potential [9].

Biochar physico-chemical properties vary to a large extent depending on a multitude of factors, especially the reacting conditions at which it is prepared [11]. The operating parameters that have been reported to influence its properties are the heating rate, the highest operating temperature, the pressure, the reaction residence time, the reactor design, the biomass pretreatments, the flow rate of inputs and the post treatments [12]. Among these numerous factors, temperature is considered one of the most important affecting char properties as it provides the necessary activation energy of the implied reactions, furthermore it influences physical changes as volatilization and structural melts. Studies on wood showed that the three main biomass components have been reported to decompose at temperature ranging from 200–325°C for hemicelluloses, from 240 to 375°C for cellulose and from 280 to 500°C for lignin [8, 12]. Demirbas [13] studied carbonization (medium rate pyrolysis at 10°C/s) for several agricultural and forest residues, for temperatures ranging from 300 to 900°C. For every feedstock, solid products were found to decrease, gases increased and liquid fractions decreased, furthermore, higher heating values (HHV) were also found to increase for increasing temperature.

Pyrolysis gaseous environment also plays a major role. Conventionally, pyrolysis operates within an environment free of oxygen, therefore using inert gases such as nitrogen. CO_2 is also an inert gas at room temperature, however, increasing temperature, CO_2 becomes reactive and tends to react with carbon char through the reverse-Boudouard reaction:

$$C + CO_2 \rightarrow 2\ CO \qquad (1)$$

There are numerous thermochemical processes that could take beneficial advantage of using CO_2. In combustion field, especially in coal industry, the Oxyfuel Technology which consists of using flue gases with pure oxygen (CO_2, H_2O and O_2) instead of air (N_2, O_2, CO_2 and H_2O) appears as a promising technology especially from CO_2 output purity and sequestration point of view [14]. In addition, use of flue gases as a gas source contains residual heat which is beneficial for energy flows within the overall process [15]. In gasification, use of CO_2 to enhance Boudouard reaction and CO formation for syngas production is also a common practice. In physical activation of char for activated carbon, CO_2 is the medium used for surface area and pore enhancement. Physical activation is usually done at temperature around 900 to 1200°C [16]. In the presence of CO_2, char yield is usually reduced towards formation of volatiles [15, 17]. Jindarom *et al.* [16] observed that for temperatures around 350 to 750°C, the use of CO_2 in comparison of N_2 resulted in chars with higher char specific surface and basicity levels.

High specific surfaces as well as high porosity are two desired characteristics for soil amendment applications since it may represent a favourable site for beneficial microorganisms' proliferation as well as an exchange surface for nutrient retention or water absorption attributes [1]. Under pyrolysis conditions, specific surface was observed to reach peak values at temperatures varying from 350 to 700°C depending upon a few factors, such as biomass types [2, 18]. Decrease in specific surface area at higher temperature was attributed, by some researchers, to melts in char structure. Increasing retention time was also shown to increase porosity since it allowed most volatilization reactions to reach completion. As for the heating rates, lower heating rates would allow volatile release with less morphological changes, therefore maintained natural porosity in comparison with higher rates which would even lead to morphological transformation of cell structure [12].

In this research, char was prepared from switchgrass (*Panicum virgatum*). Switchgrass was chosen due to its beneficial characteristics as an energy crop such as its hardiness in harsh climate conditions and poorer soils [19, 20]. In a first step, a homemade fixed bed reactor was tested under specific conditions simulating a high heating rate reactor operating under N_2 atmosphere. In a second step, the same reactor was then operated at lower heating rate, but adding new atmosphere conditions, first under N_2 and secondly under CO_2. Temperature, heating rate and gas atmosphere were the two main parameters studied and their effect on char physico-chemical characteristics was analysed. The char obtained was characterized for proximate and ultimate analyses, for BET as well as for its calorific value. Some of the chars organic content compositions were verified by Soxhlet extractions using dichloromethane and extracts were analysed by GC-MS. Bio-oil yield was also verified and the ones from slow heating rate were analyzed by GC-MS.

2 Materials and method

2.1 Biomass

Switchgrass (*panicum virgatum*) Cave-in-Rock species utilized for the experiments was grown in the Eastern Township region of Quebec, Canada. The crop was grown in a clay soil and had a 1.5 years root system. The feedstock was harvested in Fall (November 2009) at moisture content below 10% in 500 kg bales and stored in open structures protected from rain. Switchgrass samples used for experiments were maintained in unsealed plastic bags at room temperature (20–25°C).

2.2 Experimental setup

For experimentations, a lab scale fixed bed pyrolysis reactor was used. It consists of a stainless steel (SS) cylindrical reactor of 26.6 mm ID and 0.44 m length installed in a 2400 W electric tube furnace mounted vertically. Biomass is added to the reactor using a biomass carrier, fig. 1. The latter is 13 cm long, 25 mm OD in diameter and is made of stainless steel. Carrier's top, bottom and along the length is partially opened in order to allow gas flow through its structure. A wire mesh is installed within the carrier to retain the biomass from falling down. The reactor's entrance is controlled via a gate valve equipped with a motor for a quick control.

During the initial experiments operating at higher heating rate, a preheater installed underneath the reactor was used in order to preheat the biomass carrier at reactor's temperature. The carrier has a top opening in order to fill it with biomass. The action of filling the reactor and taking it into the reactor is done for every test in less than 10 s. During the second step of experiments, at lower heating rates, a preheater for biomass carrier was not used. The biomass carrier is attached to a thermocouple (T/C), which serves as a rod to move it into the reactor. The T/C also provides an indication on temperature progress within biomass carrier.

Nitrogen and carbon dioxide are the two gas vectors utilized while also serve to enhance heat and mass transfer between gases and feedstock. Gas flow input is maintained at 0.115 L/s (STP: 101.3kPa – and 25°C) for every test and is controlled by a flowmeter before entering the reactor. The flowmeter is calibrated with respect to each gas. Nitrogen gas and CO_2 are respectively preheated before entering the reactor using an inline gas heater and while passing in the annulus region between reactor and tube furnace. Gases and volatiles exiting the reactor pass by a heated line maintained around 200 ±25°C before being diverted to an exhaust system.

For pyrolysis vapours' mass balance, vapours are collected within two consecutive glass condensers; both are kept within isothermal container filled with dry ice. Pyrolysis gases are obtained by difference from char and condensable liquid yields.

Real time temperature monitoring is done on the overall system using National Instrument Data Acquisition System NI cDAQ-9172 equipped with Labview software. Thermocouples are positioned all over the system as shown in fig. 1. In fig. 1, T corresponds to thermocouples, where T1 refers to biomass carrier, T2 to gas outlet, T3, T4 and T5 to annulus region between tube furnace and reactor, T6-T9 to outlet gas line, inside and at surface temperature and T10 at gas inlet to monitor gas inlet temperature. Pressure (P1, fig. 1) is verified at reactor exit using a mechanical pressure gauge in order to maintain an absolute pressure close to 101.3 kPa in the reactor.

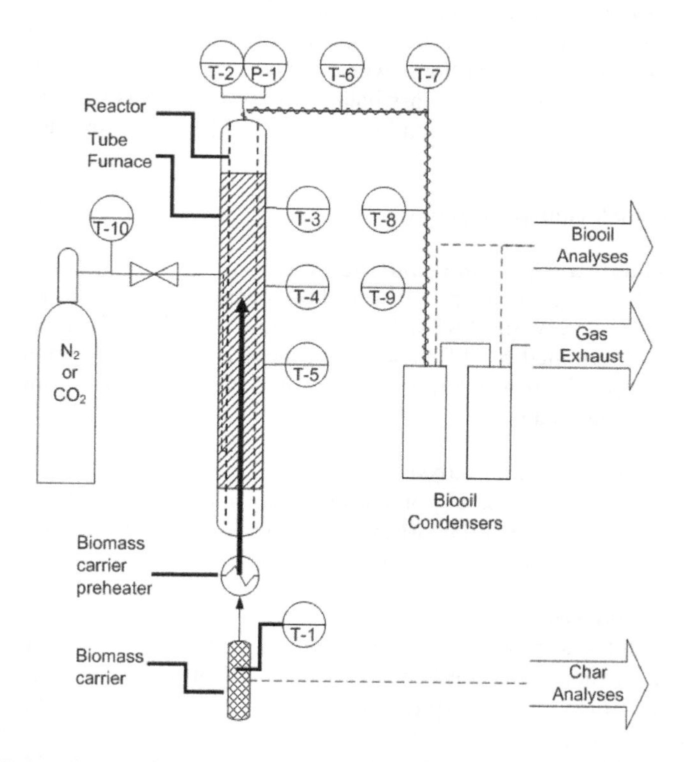

Figure 1: Schematic representation of the experimental setup.

2.3 Biomass char production and products mass balance

Biomass was first cut into pieces shorter than 10 cm and humidity level, which ranged between 4 and 10%, was monitored all along the char production periods. Typically one gram of switchgrass was inserted in the biomass carrier per batch. For each batch operating at higher heating rates, the biomass carrier was first preheated until reaching reactor's temperature, before being rapidly filled with biomass and inserted in the core of the reactor. This resulted in an almost instantaneous heat up of biomass. Gas environment used at higher heating rate is N_2. At lower heating rate, a comparison between N_2 and CO_2 was made. In these latter conditions, the biomass carrier was not preheated, biomass was inserted in the carrier, approached to reactor's opening to flush air and rapidly entered within the reactor. Biomass is kept in reactor for a desired residence time of 2.5 minutes, after what it was removed from the reactor, emptied, its content cooled down within a desiccator and transferred within hermetic containers until further analysis. During experiments at high heating rate, time was monitored from the moment the biomass was inserted in the preheated carrier. For moderate heating rates, time was counted once the temperature reading indicated the desired temperature. It took about 7.5 and 9.5 min. to reach 500 and 300°C respectively for slow heating rates conditions. For all batches, tube furnace was maintained at a fixed temperature. Chars were produced at 300 and 500 ± 10°C. Chars produced for specific conditions were mixed together before analysis. For moderated heating rate only, a mass balance on products was performed. Bio-oil was collected using condensers and non-condensable gases were determined by difference.

2.4 Feedstock characterization

2.4.1 Proximate analysis and calorific value

Char as well as original feedstock were both characterized for ash content and volatile content following ASTM E1755-01 and ASTM E 872 methods. Fixed carbon was determined by difference from the ash and volatile content. Higher heating value (HHV) was determined using an oxygen bomb calorimeter; Model 1341 from Parr company.

2.4.2 Organic elemental analysis

For CHON organic elemental content analysis, finely powdered feedstock was analyzed using a Leco TruSpec Micro CHNS equipped with the TruSpec Micro Oxygen Add-On Module for oxygen analysis.

2.4.3 BET analyses

Biomass samples were first sieved and the range between 40 and 60 mesh was used for BET analyses. Surface area characterization was performed using Micromeritic physical adsorption apparatus, ASAP 2020, using nitrogen adsorption based on BET theory. Degas conditions were done under vacuum from an initial heating rate of 10°C/min up to a temperature of 110°C kept for a minimum of 16 hours.

2.4.4 Organic components, extractions and GC-MS analysis

Char organic components extraction was done using Soxhlet. These tests were conducted on high heating rate chars only. Prior to Soxhlet extractions, char was grinded to a fine powder. For each extraction, one gram of char was placed in Whatman cellulose cartridges (25 mm ID × 90 mm) and extracted with 150 ml of dichloromethane. The solution was concentrated using roto-evaporator. Biooils obtained from slow heating rate conditions were also analyzed. These latter collected into condensers were diluted into acetone before being injected into GC-MS. The gas chromatogram utilized is a HP 5890 Series II equipped with HP 5971A Mass Selective Detector and HP 7673 Controller.

3 Results and discussion

3.1 Heating rate effect

At 300°C, a change in heating rate does not seem to have a major effect on physical properties despite the fact the char yield varied slightly as well as higher heating value (HHV), table 1. Char yield would have been expected to decrease with increased heating rate, as well as volatile content since generally increasing heating rate results in higher liquid yields and lower char yields, as for 500°C. Perhaps that at lower temperature, this principle does not hold and the longer temperature progression temperature period had a more pronounced effect on char yield. Char yield being lower at 300°C is at least consistent with the higher fixed carbon content that may have resulted from longer residence time, releasing more volatiles at temperatures lower than 300°C. This latter would also explain the increase in heating value content. Higher heating value obtained for 500 °C in this research is really close to what was previously reported in literature using a bench scale fluidized bed reactor design (21.5 vs. 18.6 MJ/kg) [21]. At 500°C, decreasing heating rate seems to have a pronounced effect on surface area passing from 52 to 82 m^2/g. Effect on biooil content was not observed since biomass carrier preheating step for higher heating rates resulted in some volatiles losses. It has to be kept in mind that oxygen introduction resulting from carrier preheating may have major effect on the results.

3.2 Gas environment effect

At 300°C, gas environment does not seem to have a major effect on physical properties. All parameters studied did not vary, which is consistent to theory mentioning that CO_2 is inert at lower temperature [16] (table 1). On the other hand, despite the fact it would require further analyses, presence of compounds such as syringol, methoxyeugenol and palmitic acid was noted in 300°C CO_2 oils which was not at 300°C N_2. A further investigation on oil is however required in order to prove that observation.

At 500°C though, gas environment seems to have a more pronounced effect. First of all, decrease in char yield and carbon content in presence of CO_2 shows that Boudouard reversed reaction could have taken place, resulting in higher carbon conversion and gas yield. Since 500°C is a relatively low temperature for Boudouard reaction to take place, it seems that at least an element acted as a catalyst. It this case, two factors may be the catalyst source: the first could be Ni element present in stainless steel of the reactor, the thermocouple and the biomass carrier. As reported by Osaki and Mori [22], in presence of Ni, the reversed Boudouard reaction along gasification operates at same rate at 500°C than without catalyst at 800°C. The second factor could come from the high ash content into switchgrass that could act as catalyst for Boudouard reversed reaction. Butterman and Castaldi [15] observed that activation energy for pyrolysis and gasification in CO_2 atmosphere was lowered in presence of herbaceous feedstock in comparison to wood. This difference was attributed mostly to higher ash content in herbaceous feedstock than for wood.

Specific surface is also another parameter which was influenced by carbon dioxide. At 300 °C there was no change noticed, however at 500°C, specific surface passed from 82 to 189 m^2/g, (table 1), which is consistent to results from Jindarom *et al.* [16] who observed a drastic change in specific surface from 550°C.

3.3 Temperature effect

Temperature treatments add a pronounced effect on surface area. Increasing temperature from 300 to 500°C increased surface area from less than 1 m^2/g for all 300°C char production studied, to more than 50 m^2/g at high heating rate and to more than 80 m^2/g at low heating rate, table 1. Despite the fact the high increase in surface area at 500°C might be beneficial on a soil amendment application, the decrease in char yield passing from about 68% to 14% must be considered among the economics of the process depending on the application. Char yield from 66 to 70% at 300°C could result mostly from hemicellulose decomposition as reported by Prins *et al.* [8]. This latter assumption would be consistent with switchgrass hemicelluloses content which was estimated at 36.3%, table 2. In each case, char yield decreased with increasing temperature with respect to specific factors studied. This mass balance is consistent with observations made by other authors showing that cellulose decomposition occurs at temperature from 305 to 375°C and lignin at temperature from 250 to 500°C [8]. Switchgrass cellulose content represents 33.7% of biomass composition (table 2), lignin being more reluctant to decomposition, this latter would then potentially balance the remaining losses. Boateng *et al.* [21] worked on the same species using a bench-scale fluidized bed fast pyrolysis reactor. They run their tests at a temperature of 500°C and the char produced corresponded to 12.9% of the original biomass, which is close to the actual values obtained [21].

It has to be noted that char was removed from the reactor once the residence time was completed. As a result, for higher temperature, 500°C, the biomass and biomass carrier remain at temperature high enough for char reacting with the ambient oxygen, resulting in char glowing red. This definitively may have an effect on char yields by reducing it as well as an effect on its characteristics.

3.4 DCM extractions and analyses

Char organic content using Soxhlet extractions with dichloromethane were used to investigate if PAHs would be present within the char studied. This study was completed only for the high heating rate and within the studied conditions, only naphthalene was qualitatively noticed at 500°C temperature. In literature, operating temperature increase was already reported to increase PAH formation

Table 1: Torrefaction and pyrolysis products analyses.

Production conditions				Mass balance			Char analyses								
Feedstock	**Heating rate**	**T° [°C]**	**Gas Env.**	**Char [%]**	**Oil [%]**	**Gas [%]**	**C [%]**	**H [%]**	**O [%]**	**N [%]**	**Ash [%]**	**Volatiles [%]**	**FC [%]**	**HHV MJ/kg**	**S.A. m^2/g**
Raw sg.	n.a.	n.a.	n.a.	n.a.	n.a.	n.a.	44.5	5.8	45.7	<1	3.7	81	15.3	19.5	–
Char	FAST	300	N_2	70.9	–	–	51.3	5.5	38.8	<1	6.7	72.8	20.5	20.3	<1
Char	FAST	500	N_2	12.4	–	–	63.9	2.6	15	<1	26.6	20.8	52.6	19.6	52
Char	SLOW	300	N_2	66.9	9	24.1	53.3	5.5	38.9	<1	4.7	70.9	24.4	23.5	<1
Char	SLOW	300	CO_2	66.6	10.6	22.8	54.9	5.5	38.4	<1	5.4	68.9	25.7	23.1	<1
Char	SLOW	500	N_2	15.9	10.6	73.6	65.4	2.6	17.4	1	15.1	19.7	65.2	23.5	82
Char	SLOW	500	CO_2	13.6	6.9	79.5	61.8	2.1	18.6	1	14.7	21.5	63.8	25.1	189

n.b.: Percentages are expressed on a dry basis.
Legend: sg.: Switch grass; T: Temperature; FC: Fixed carbon; n.a.: Not applicable; Env.: Environment; S.A.: Surface area.

Table 2: Original composition of *Panicum virgatum* straw.

Identification	Standard test	Mass Units (± *S.D*)*
Ash	ASTM E1755-01	3.7 (± *0.2*)
α-cellulose	ASTM D1103-60	33.7 (± *0.7*)
Hemicellulose	By difference	36.3 (± *1*)
Lignin	ASTM D1106-56	17.3 (± *0.6*)
Total		**100**

*Expressed in terms of 100 mass units of oven dry switchgrass straw.

[23, 24]. Since extractions requires substantial amount of char in order to be extracted, further investigation of such parameter will be carried using a reactor with higher feedstock capacity.

4 Conclusion

Preliminary investigations of char formation from switchgrass show results similar to those reported so far. Lowering the heating rate would enhance the specific char surface area especially at 500°C. It would also result in slightly higher char yield, which may be beneficial from a production point of view. Carbon dioxide showed to have a great impact on specific surface at 500°C, leading to a specific surface of 189 m^2/g, which more than doubled the N_2 results. For a biochar-soil amendment application, where optimized yields are desired as well as high specific surface, residual CO_2 might then represent a strong potential since the yield compared to 500°C N_2 was reported as similar. Leaching of ashes prior to pyrolysis or using a reactor free of Ni could be recommended in further experiments in order to target whether an element or another was responsible for the enhancement of CO_2 effect. Despite the fact CO_2 seems to be inert at 300°C based on physical characteristics of chars, preliminary results for biooil content analysis revealed some differences which must be further investigated. Further investigation about chemical composition of char obtained under 500°C would also be interesting since the appearance of PAHs occurred at higher heating rates with N_2, however in presence of CO_2, the content could potentially be reduced since char tar content could react with CO_2 and enhance CO formation. It was also reported that final reacting conditions at open atmosphere resulted in decrease of char yield. Despite the fact this obvious reaction with oxygen could easily be reduced by waiting for char to cool down, the effect of such post treatment on char properties should also be analyzed further with respect to char value additions and the practical applications of the technology.

Acknowledgements

The authors would like to thank Le Fonds québécois de la recherche sur la nature et les technologies (FQRNT) as well as the Université de Sherbrooke, Faculty of Engineering and Department of Chemical Engineering, for the financial support for Guillaume Pilon's PhD thesis; the Université de Sherbrooke, the Faculty of Engineering and the Department of Chemical Engineering and Biotechnological Engineering for supporting this research (through starting funds for Jean-Michel Lavoie and lending research equipment) as well as the Industrial Chair in Ethanol Cellulosic in lending research equipment. The authors would also like to acknowledge Ms. Eva Capek, Mr. Henri Gauvin, Mr. Serge Gagnon, Mr. Marc G. Couture and Mr. Michel Trottier of the Université de Sherbrooke for their technical support. Finally, Mr. Daniel Clément, switchgrass producer from the Eastern Township, for providing the feedstock for the experiments.

References

[1] Lehmann, J. & Joseph, S., *Biochar for Environmental Management*, Earthscan: London – Washington, DC, 2009.

[2] McLaughlin, H., Anderson, P.S., Shields, F.E. & Reed, T.B., All biochars are not created equal, and how to tell them apart. *North American Biochar Conference*, 2009.

[3] Gaunt, J.L. & Lehmann, J., Energy balance and emissions associated with biochar sequestration and pyrolysis bioenergy production. *Environmental Science & Technology*, **42(11)**, pp. 4152–4158, 2008.

[4] Lehmann, J., A handful of carbon. *Nature*, **447(7141)**, pp. 143–144, 2007.

[5] Laird, D.A., The charcoal vision: a win win win scenario for simultaneously producing bioenergy, permanently sequestering carbon, while improving soil and water quality. *Agronomy Journal*, **100(1)**, pp. 178–181, 2008.

[6] Boateng, A.A., Characterization and thermal conversion of charcoal derived from fluidized-bed fast pyrolysis oil production of switchgrass. *Industrial and Engineering Chemistry Research*, **46(26)**, pp. 8857–8862, 2007.

[7] Uslu, A., Faaij, A.P.C. & Bergman, P.C.A., Pre-treatment technologies, and their effect on international bioenergy supply chain logistics. Techno-economic evaluation of torrefaction, fast pyrolysis and pelletisation. *Energy*, **33(8)**, pp. 1206–1223, 2008.

[8] Prins, M.J., Ptasinski, K.J. & Janssen, F.J.J.G., Torrefaction of wood: Part 1. Weight loss kinetics. *Journal of Analytical and Applied Pyrolysis*, **77(1)**, pp. 28–34, 2006.

[9] Brewer, C.E., Schmidt-Rohr, K., Satrio, J.A. & Brown, R.C., Characterization of biochar from fast pyrolysis and gasification systems. *Environmental Progress & Sustainable Energy*, **28(3)**, pp. 386–396, 2009.

[10] Mohan, D., Pittman, C.U. & Steele, P.H., Pyrolysis of wood/biomass for bio-oil: a critical review. *Energy & Fuel*, **20(3)**, pp. 848–889, 2006.

[11] Joseph, S., Peacocke, C., Lehmann, J. & Munroe, P., Developing a biochar classification and tests methods (Chapter 7). *Biochar for Environmental Management – Science and Technology*, ed. J. Lehmann & S. Joseph, Earthscan: London and Washington DC, pp. 107–112, 2009.

[12] Downie, A., Crosky, A. & Munroe, P., Physical Properties of Biochar, *Biochar for Environmental Management – Science and Technology*. ed. J. Lehmann & S. Joseph, Earthscan: London and Washington DC, pp. 13–29, 2009.

[13] Demirbas, A., Carbonization ranking of selected biomass for charcoal, liquid and gaseous products. *Energy Conversion and Management*, **42(10)**, pp. 1229–1238, 2001.

[14] Buhre, B.J.P., Elliott, L.K., Sheng, C.D., Gupta, R.P. & Wall, T.F., Oxy-fuel combustion technology for coal-fired power generation. *Progress in Energy and Combustion Science*, **31(4)**, pp. 283–307, 2005.

[15] Butterman, H.C. & Castaldi, M.J., Biomass to fuels: impact of reaction medium and heating rate. *Environmental Engineering Science*, **27(7)**, pp. 539–555, 2010.

[16] Jindarom, C., Meeyoo, V., Kitiyanan, B., Rirksomboon, T. & Rangsunvigit, P., Surface characterization and dye adsorptive capacities of char obtained from pyrolysis/gasification of sewage sludge. *Chemical Engineering Journal*, **133(1–3)**, pp. 239–246, 2007.

[17] Minkova, V., Marinov, S.P., Zanzi, R., Björnbom, E., Budinova, T., Stefanova, M. & Lakov, L., Thermochemical treatment of biomass in a flow of steam or in a mixture of steam and carbon dioxide. *Fuel Processing Technology*, **62(1)**, pp. 45–52, 2000.

[18] Sharma, R.K., Wooten, J.B., Baliga, V.L., Lin, X., Geoffrey Chan, W. & Hajaligol, M.R., Characterization of chars from pyrolysis of lignin. *Fuel*, **83(11–12)**, pp. 1469–1482, 2004.

[19] He, R., Ye, X.P., English, B.C. & Satrio, J.A., Influence of pyrolysis condition on switchgrass bio-oil yield and physicochemical properties. *Bioresource Technology*, **100(21)**, pp. 5305–5311, 2009.

[20] Martel, H. & Perron, M.-H., Compilation des essais de Panic Érigé réalisés au Québec. Centre de référence en agriculture et agroalimentaire du Québec, 2008. http://www.craaq.qc.ca/data/DOCUMENTS/EVC026.pdf

[21] Boateng, A.A., Daugaard, D.E., Goldberg, N.M. & Hicks, K.B., Bench-Scale fluidized-bed pyrolysis of switchgrass for bio-oil production. *Industrial & Engineering Chemistry Research*, **46(7)**, pp. 1891–1897, 2007.

[22] Osaki, T., & Mori, T., Kinetics of the reverse-Boudouard reaction over supported nickel catalysts. *Reaction Kinetics and Catalysis Letters*, **89(2)**, pp. 333–339, 2006.

[23] McGrath, T., Sharma, R. & Hajaligol, M., An experimental investigation into the formation of polycyclic-aromatic hydrocarbons (PAH) from pyrolysis of biomass materials. *Fuel*, **80(12)**, pp. 1787–1797, 2001.

[24] Nakajima, D., Nagame, S., Kuramochi, H., Sugita, K., Kageyama, S., Shiozaki, T., Takemura, T., Shiraishi, F. & Goto, S., Polycyclic aromatic hydrocarbon generation behaviour in the process of carbonization of wood. *Bulletin of Environmental Contamination and Toxicology*, **79(2)**, pp. 221–225, 2007.

The heterogeneous reaction between tar and ash from waste biomass pyrolysis and gasification

Q. Wang, T. Endo, P. Apaer, L. Gui, Q. Chen, N. Mitsumura, Q. Qian, H. Niida, S. Animesh & K. Sekiguchi
Department of Environmental Science and Technology, Graduate School of Science and Engineering, Saitama University, Japan

Abstract

Fossil energy resources that are available in the world are exhaustible. Therefore, renewable biomass has attracted a lot of attention as a future energy resource. In addition, it is an advantage that the biomass grows while absorbing CO_2, contributing to prevention of global warming. Biomass utilization technologies are classified as pyrolysis and gasification, fermentation and combustion. Fuel gases and synthesis gases produced by the pyrolysis and gasification are used for power generation, heating and as chemical products. However, pyrolysis and gasification processes also generate condensable organic compounds, so-called "tar". Most of the tar contents are present as gases at high temperature. However, when they are cooled down at temperatures lower than their boiling point, causing a black oily liquid that leads to equipment failure, appropriate processing is required. As a processing method, the use of catalytic tar decomposition has been widely studied. In the work presented here, thermal decomposition of cellulose was carried out in an experimental apparatus modeling a fluidized bed gasifier. During thermal decomposition of cellulose, tar and gas are generated, tar is collected and cooled, and the gases are measured by a gas-chromatograph with a flame ionization detector (GC-FID) and with a thermal conductivity detector (GC-TCD). Then, K and Ca are selected as the catalysts of alkali metals and alkaline earth metals contained in the waste biomass. They are present in the state of oxide or carbonate during pyrolysis and gasification. A similar experiment was conducted. The amount of condensable products and heavy tar were decreased by installing K_2CO_3 and $Ca(OH)_2$. Additionally, they brought a further gas production. It can be concluded that alkali metal compound (K_2CO_3)

and alkaline earth compound (CaO) have a catalytic effect to decompose tar contents, to enhance gaseous production.
Keywords: biomass, pyrolysis, heterogeneous reaction, alkali metal, alkaline earth metal, tar decomposition.

1 Introduction

Focusing on developing countries, it is expected that the amount of energy consumption is increased in the world. Oil, coal and natural gas are exhaustible resources used to fulfil the energy requirements in the world. Large amounts of carbon dioxide emission occur when these energy sources are used and this promotes global warming. Recently, biomass has attracted attention as a renewable energy resource. Thermal decomposition and gasification are methods used to convert biomass to energy. The gas obtained by the thermal decomposition and gasification is mainly composed of H_2 and CO. From these synthesis ammonia, liquid fuel, methanol and chemical products as well as a variety of derivative products are manufactured [1]. However, an amount of condensable organic compound called tar is produced during thermal decomposition and gasification. The tar clogs the pipe of gasifier and breaks the turbines. Both mechanical methods and thermal cracking have been proposed to remove the tar.

Mechanical methods do not allow energy recovering from tar which is just removed from gaseous products, while thermal cracking requires high temperature (>1100°C) to convert them [2]. Catalytic tar decomposition has been proposed to overcome these drawbacks. In addition, the use of a catalyst can enhance gas formation and modify the gaseous composition promoting the reforming reactions of hydrocarbons. Tar removal using a catalyst has been extensively studied. It is reported that nickel-based catalysts, alkaline metal oxides and alkaline earth metal oxides are suitable to reduce the amount of tar [3]. Alkali metal and alkaline earth metal are present in the ash component of the biomass.

The focus here is on tar reduction in a fluidized bed gasifier such as a fluidized bed and spouted bed system. Even though experiments on thermal decomposition and gasification of biomass in a fixed bed system have been widely studied, studies on the heterogeneous reaction between tar and ash are yet to be performed [4]. Using an experimental system that assumes the heterogeneous reaction of ash and tar in a fluidized bed gasifier, the evaluation of the catalytic effect of the ash is significant. In this study, an experimental thermal decomposition of cellulose has been conducted by constructing an experimental device that assumes the heterogeneous reaction in a fluidized bed gasifier. Cellulose is a major component of wood and plant biomass [5]. By introducing the ash in the experimental thermal decomposition of cellulose, the effect of ash on the heterogeneous reaction with tar has been considered. The yields of char from the results of the thermogravimetry-differential thermal analysis (TG-DTA) experiment have been calculated; the yields of condensable products cooled down and collected in the test tube have been measured; the yields of gas by GC-TCD/FID have also been measured.

2 Materials and experimental methods

2.1 Materials

The tested experimental sample was microcrystalline cellulose with an average diameter of 50 μm (SERVA). The chemical formula for this sample polymer could be approximated as $(C_6H_6O_5)_n$, confirming the linearly polymerized structure of the glucopyranose linked by β-1, 4-glycosidic bonds. The analysis of the cellulose had been performed using a CHN corder (Model MT-5, Yanaco Co. Ltd., Japan) for elemental analysis and proximate analysis, which were both performed. The cellulose composition is presented in table 1.

Table 1: Elemental analysis and proximate analysis of cellulose.

Elemental analysis (wt.%)				Proximate analysis (wt.%)			
C	H	N	O	Moisture	Volatiles	Fixed carbon	Ash
43.31	6.23	0.00	50.46	7.18	86.90	5.92	0.00

Potassium and calcium were alkali metal and alkaline earth metal species contained in the biomass. K_2CO_3 (Wako, assay min. 99.5%) and $Ca(OH)_2$ (Wako, assay min. 96.0%) were used in the study as ash model. Silicon dioxide (Wako) was used as a fluidizing medium [6].

2.2 TG-DTA experiment

The pyrolysis of cellulose was carried out in a TG-DTA (Model DTG-60, Shimadzu Co. Ltd., Japan), in order to calculate the yield of char and survey the pyrolysis behavior. The sample was weighed 5.00–7.00 mg and placed on the scales in the apparatus. The sample was heated up to 900°C at a constant heating rate of 10°C/min. Argon at a flow rate of 70 ml/min was used as the carrier gas to provide an inert atmosphere for pyrolysis and to remove the gaseous and condensable products.

2.3 Thermal decomposition of cellulose

2.3.1 Experimental apparatus for cellulose pyrolysis and gasification

Figure 1 shows the scheme of the experimental setup used for cellulose pyrolysis and gasification. It was composed of a gas feeding system, a pyrolysis system, a tar decomposition system, a condensable products trapping system and a gaseous products measurement system. The pyrolysis and tar decomposition systems consisted of connecting the two stainless reactors (I.D. 21.4 mm, length 500 mm). Cellulose, K_2CO_3, $Ca(OH)_2$ and silicon dioxide were placed on the mesh (40 μm) in each reactor. This apparatus can be independently heated in two different electric furnaces while the gas mixture is coming to contact with the catalyst. Therefore, a heterogeneous reaction between tar and ash can be carried out using the apparatus. The lines between the first and second reactor as well as between the second reactor and the condensable products trapping system were heated at between 300 and 400°C to avoid the condensation of tar. The condensable products were collected by cooling the test tube and the collection efficiency was improved by using glass beads. The Cooling bath was kept below –3°C by mixing water, ice and sodium chloride. The Gaseous products were measured by a GC-TCD/FID (Model GC-2014, Shimadzu Co. Ltd., Japan). H_2 and CO, CH_4, CO_2 were measured by a GC-TCD while hydrocarbons (C_2H_6 and C_2H_4, C_3H_8, C_3H_6, iso-C_4H_{10}, n-C_4H_{10}) were measured by a GC-FID.

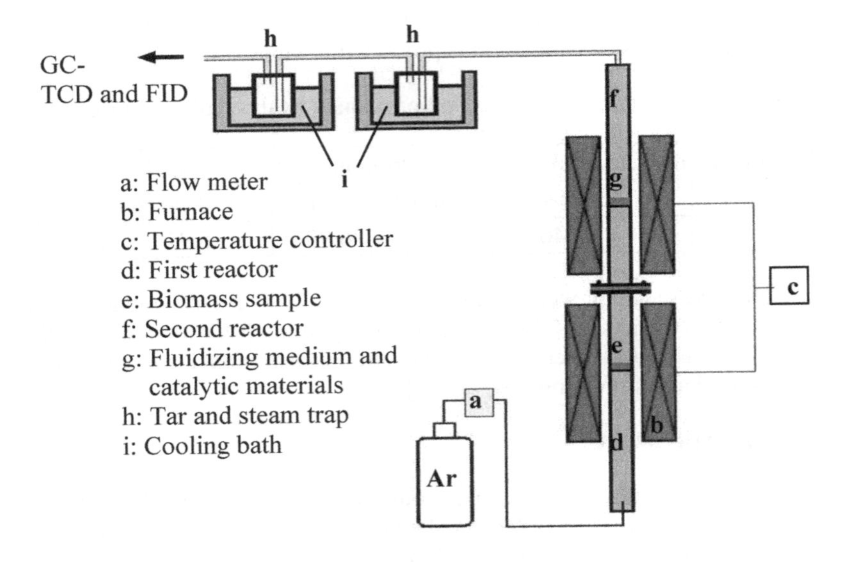

Figure 1: Experimental setup for cellulose pyrolysis and gasification.

2.3.2 Experimental procedure

The operative conditions are reported in table 2. The cellulose, catalysts and silicon dioxide were placed in each reactor under the respective conditions of table 2. The second reactor was heated up to a predetermined temperature under argon at constant flow rate. Then, the first reactor was heated up (at 10°C/min up to 900°C) and the gaseous products were measured at every 80°C to 900°C from 180°C by a GC-TCD/FID.

2.3.3 Yields of pyrolysis products

The yields of char were calculated from TG-DTA analysis. The yields of condensable products were determined by the weight difference of the test tube before and after the experiment. The yields of gaseous products were calculated from the measurement results of GC-TCD/FID. Also, the amount obtained by subtracting the amount of char, the amount of condensable products and the amount of gaseous products from the amount of cellulose was evaluated as a heavy tar. The heavy tar is a tar which remains in the apparatus.

Table 2: Operative conditions used for tests.

Item	First reactor	Second reactor		
		none	K_2CO_3	$Ca(OH)_2$
Cellulose amount (g)	2.00	-		
Catalyst amount (g)	-	-	0.691	0.741
Silicon dioxide amount (g)	-	5.00	4.309	4.259
Heating rate (°C/min)	10	-		
Maximum temperature (°C)	900	500, 700, 800, 900	700, 800	700, 800
Argon flow rate (ml/min)	70			

3 Results and discussion

3.1 Results from TG-DTA analysis

The pyrolysis characteristics, both TG (wt.%) and DTA (μV) curves of cellulose determined with a TG-DTA are shown in fig. 2. Rapid weight loss of cellulose was observed between 300 and 400°C and then the weight loss had progressed slowly. Endothermic peak was observed between 300 and 400°C, indicating that the thermal decomposition of cellulose was occurring within this temperature range [7]. Furthermore, the weight loss was also observed in the same temperature range. Thus, it should be predicted that a large amount of thermal

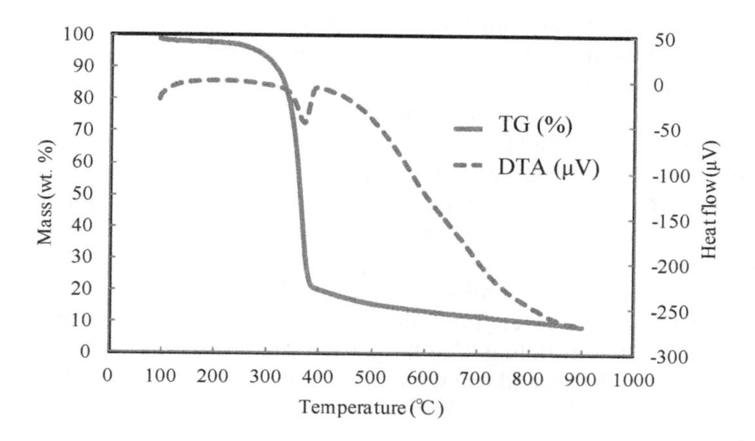

Figure 2: Pyrolysis curves of cellulose in the TG-DTA analysis.

decomposition products are released at this temperature. As a gradual weight loss continued above 400°C, it should be predicted that a small amount of thermal decomposition products would have been released. Experiments were carried out several times under the same conditions and the measurements were within ± 5% maximum error. Thus, in this study, char yields of cellulose pyrolysis were calculated to be 8.90% (wt.%).

3.2 Thermal decomposition of cellulose

Figures 3 to 6 show the molar quantity (mmol/g-cellulose) of gaseous products generated by thermal decomposition of cellulose under each experimental condition. Also, table 3 shows the total molar quantity (mmol/g-cellulose) of gaseous products generated by thermal decomposition of cellulose under each experimental condition.

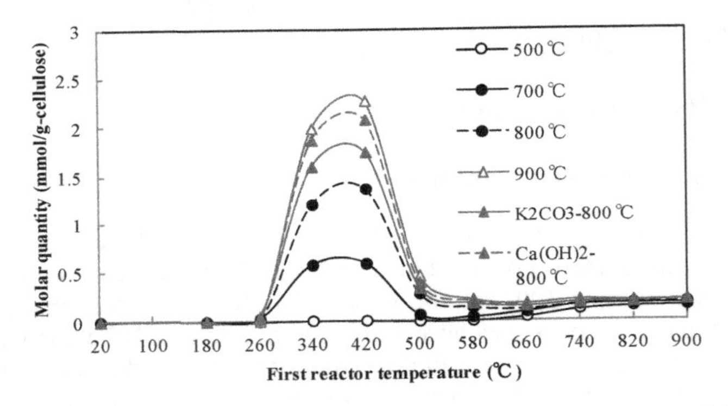

Figure 3: Molar quantity of H_2 under each second reactor conditions.

At the second reactor temperature of 900°C, the total molar quantity of H_2 was maximum while it was minimum at the second reactor temperature of 500°C. At the second reactor temperature of 500°C, the generation of H_2 began to increase rapidly above, approximately, 600°C. However, H_2 generation was remarkable in between 300 and 500°C under experimental conditions. In the case of second reactor temperature at 500°C, generation of H_2 was not observed; only a small amount was generated in that temperature range. The total molar quantity of H_2 was increased as the second reactor temperature conditions rose. The total molar quantity of H_2 was increased by approximately 7.5% after adding K_2CO_3 compared to the molar quantity observed with the presence of only silicon dioxide at 700°C. However, the total molar quantity of H_2 was increased by nearly 77.6% after installing $Ca(OH)_2$. Compared with the presence of only silicon dioxide in the second reactor at 800°C, the total molar quantity of H_2 was increased by about 28.2% and 45.5% with the presence of K_2CO_3, and $Ca(OH)_2$ respectively.

In the case of $Ca(OH)_2$ being placed in second reactor at 800°C, the total molar quantity of CO was maximum while it was minimum in the case of a second reactor temperature of 500°C. The generation of CO was significant between 300 and 500°C under all experimental conditions. Compared to the condition where only silicon dioxide was placed in the second reactor at 700°C, the total molar quantity of CO was increased by about 20.4% after adding K_2CO_3. However, it was further increased by nearly 60.4% when $Ca(OH)_2$ was installed. The quantity of CO was increased by nearly 1.2% and 51.1% with the presence of K_2CO_3 and $Ca(OH)_2$, respectively, compared to the CO quantity observed when only silicon dioxide was placed in the second reactor at 800°C. The total molar quantity of CO when $Ca(OH)_2$ was installed without changing the second reactor temperature, was higher than when the second reactor temperature was raised to 700–800°C and 800–900°C. CO was mostly generated in the pyrolysis gaseous products.

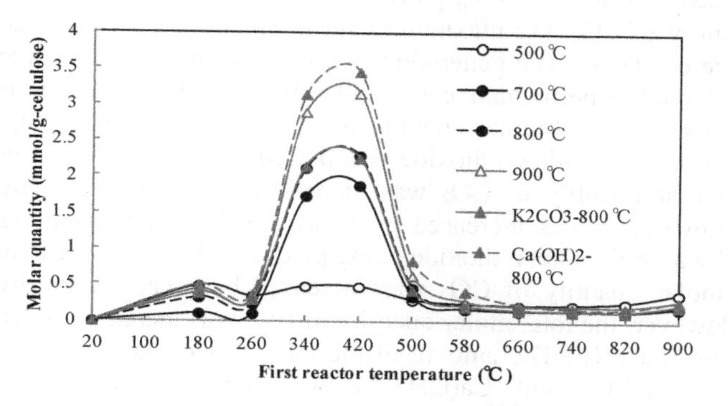

Figure 4: Molar quantity of CO under each second reactor conditions.

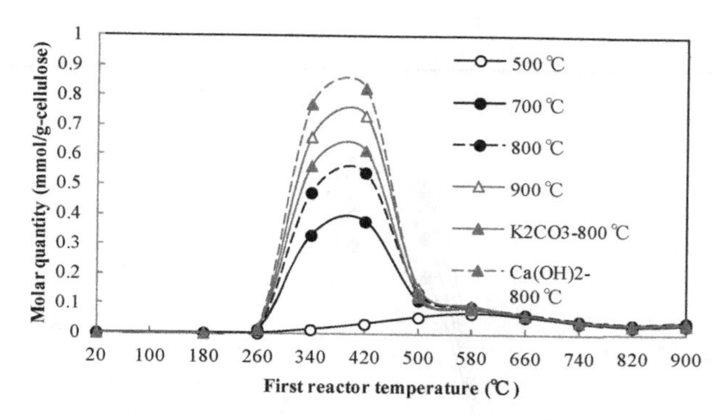

Figure 5: Molar quantity of CH_4 in each second reactor conditions.

In the case of $Ca(OH)_2$ being placed in the second reactor at 800°C, the total molar quantity of CH_4 was maximum and it was minimum at a second reactor temperature of 500°C. The generation of CH_4 was significant between 400 and 800°C when the second reactor temperature was 500°C while it was significant in between 300 and 500°C under other experimental conditions. At the second reactor temperature of 500°C, generation of CH_4 was not observed; only a small amount was generated within that temperature range. As the second reactor temperature conditions rose, the total molar quantity of CH_4 was increased. Compared to the condition where only silicon dioxide was placed in the second reactor at 700°C, the total molar quantity of CH_4 was increased by about 11.7% and 46.0% by adding K_2CO_3 and $Ca(OH)_2$, respectively. Compared with only silicon dioxide being placed in the second reactor at 800°C, the total molar quantity of CH_4 was increased by about 9.4% by installing K_2CO_3; on the other hand, the total molar quantity of CH_4 was increased by about 40.4% by installing $Ca(OH)_2$. The rate of the total molar quantity of CH_4 when $Ca(OH)_2$ was installed without changing the second reactor temperature, was higher than when the second reactor temperature was raised to 700–800°C and 800–900°C.

In the case of $Ca(OH)_2$ being placed in the second reactor at 800°C, the total molar quantity of CO_2 was maximum, and it was minimum at a second reactor temperature of 500°C. The generation of CO_2 was significant between 300 and 500°C under all experimental conditions. As the second reactor temperature condition rose, the total molar quantity of CO_2 was increased. Compared to the condition where only silicon dioxide was placed in the second reactor at 700°C, the total molar quantity of CO_2 was increased by about 39.6% by installing K_2CO_3. However it was increased by nearly 39.5% after installing $Ca(OH)_2$. Compared with only silicon dioxide being placed in the second reactor at 800°C, the total molar quantity of CO_2 was increased by about 49.4% by installing K_2CO_3. However, the total molar quantity of CO_2 was increased by about 62.0% by installing $Ca(OH)_2$. The amount of the total molar quantity of CO_2 with the presence of K_2CO_3 and $Ca(OH)_2$ without changing the second reactor temperature was higher than that generated when the second reactor temperature was raised to 800–900°C.

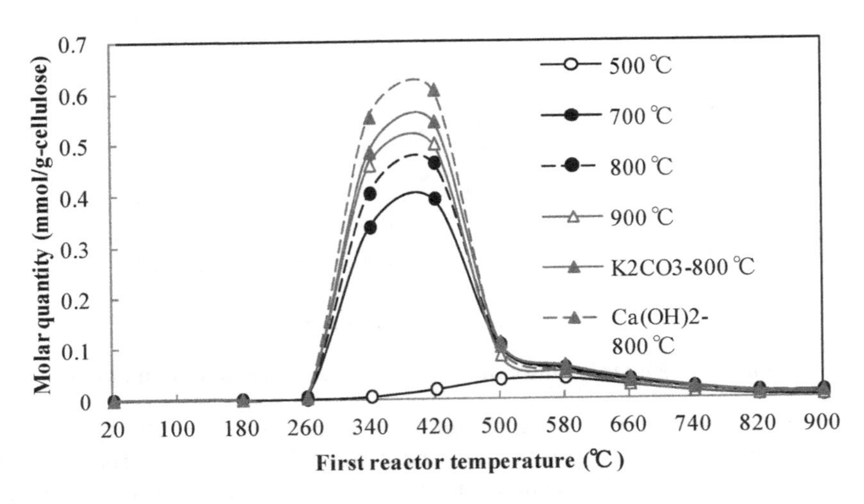

Figure 6: Molar quantity of hydrocarbons (C_2H_6 and C_2H_4, C_3H_8, C_3H_6, iso-C_4H_{10}, n-C_4H_{10}) under each second reactor conditions.

In the case of $Ca(OH)_2$ being placed in the second reactor at 800°C, the total molar quantity of hydrocarbons was maximum while a minimum quantity of hydrocarbons was observed when the second reactor temperature was 500°C. At a second reactor temperature of 500°C, the generation of hydrocarbons was significant between about 400 and 800°C. However, the generation of hydrocarbons was significant between about 300 and 500°C under the other experimental conditions. At a second reactor temperature of 500°C, generation of hydrocarbons was not observed; only a small amount occurred within that temperature range. As the second reactor temperature conditions rose, the total molar quantity of hydrocarbons was increased. The total molar quantity of hydrocarbons was increased by approximately 6.1% and 42.2% by installing K_2CO_3 and $Ca(OH)_2$, respectively, compared to the total molar quantity of hydrocarbons when only silicon dioxide was placed in the second reactor at 700°C. Compared with only silicon dioxide being placed in the second reactor at 800°C, the total molar quantity of hydrocarbons was increased by about 15.8% after installing K_2CO_3 while it was increased by about 25.4% after installing $Ca(OH)_2$. The quantity of the total molar quantity of hydrocarbons when $Ca(OH)_2$ was installed without changing the second reactor temperature, was higher than that generated when the second reactor temperature was raised to 800–900°C.

Table 3: Total molar quantity (mmol/g-cellulose) of gaseous products under each second reactor condition.

Second reactor conditions	H_2	CO	CH_4	CO_2	Hydrocarbons
500°C	0.493	3.531	0.333	0.921	0.149
700°C	1.892	5.059	1.114	1.225	0.979
700°C-K_2CO_3	2.033	6.088	1.244	1.710	1.039
700°C-$Ca(OH)_2$	3.359	8.116	1.626	1.709	1.393
800°C	3.638	6.411	1.454	1.847	1.112
800°C-K_2CO_3	4.665	6.487	1.590	2.759	1.288
800°C-$Ca(OH)_2$	5.294	9.691	2.040	2.993	1.395
900°C	5.751	8.673	1.847	2.268	1.146

The total molar quantity of all gas species was increased with the increase of the second reactor temperature; the cause of this should be a secondary pyrolysis reaction of cellulose pyrolysis products in the second reactor. At higher second reactor temperature, heavier tar could be pyrolyzed, thus, the generation of additional gas and carbon constituent is expected. Differences in temperature conditions generated a larger amount of gas from secondary pyrolysis and gasification of tar than from the primary pyrolysis. That could be the reason for the significant difference in the total molar quantity of gas production [8]. In the case of second reactor temperature conditions being the same, the total molar quantity of all gas species increased after installing K_2CO_3 and $Ca(OH)_2$. Also, under certain experimental conditions, there should be an increase in the total molar quantity of gaseous products rather than rising temperature when the catalyst was installed. On the basis of these results, it can be concluded that the addition of K_2CO_3 and $Ca(OH)_2$ significantly increased gas production. Compared to $Ca(OH)_2$, relatively higher amount of gas is produced in the presence of K_2CO_3. But the role of K_2CO_3 might have been underestimated. This is because K_2CO_3 reacts with silicon dioxide to form glass at a temperature of around 800°C [9]. Furthermore, the weight ratio of K_2CO_3 and silicon dioxide in the glass in this study is somewhat similar to that in the literature [10]. The melting temperature of K_2CO_3 is around 900°C, thus, in this study, K_2CO_3 is not decomposed into K_2O and CO_2 [11].

At a second reactor temperature of 500°C, the generation behavior of gas was very different from that under other temperature conditions and the molar quantity was also less. On the basis of these reasons, it can be estimated that secondary pyrolysis and gasification of tar did not much occurred. However, under other experimental conditions, it seems that a large amount of pyrolysis products, between 300 and 400°C, have been secondary pyrolyzed and have contributed to the active gas generation between 300 and 500°C.

Figure 7 allows a comparison of mass balance in pyrolysis of cellulose. At a second reactor temperature of 500°C, the largest fragment was condensable products while heavy tar represents the largest fraction at 700°C. On the other hand, the smallest amount of condensable products and heavy tar and total tar were found when $Ca(OH)_2$ was placed in the second reactor at 800°C. The amount of condensable products and total tar decreased, as the second reactor temperature rose. This could be due to the secondary pyrolysis reaction of cellulose pyrolysis products in the second reactor. But the amount of heavy tar in the case of a second reactor temperature of 500°C was less than that at a second reactor temperature of 700 and 800°C. It is surmised that tar composition has gradually changed to heavier one by performing pyrolysis at a high temperature, but this reaction might be insufficient at 500°C. At a second reactor temperature of 700°C, the amount of condensable products and heavy tar and total tar were

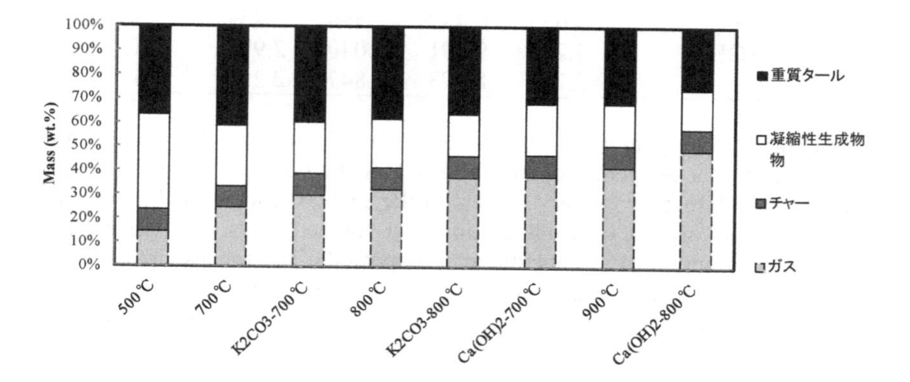

Figure 7: Comparison of mass balance on pyrolysis of cellulose.

decreased by installing K_2CO_3 and $Ca(OH)_2$. This result was the same at 800°C. On the basis of current findings, it can be concluded that K_2CO_3 and $Ca(OH)_2$have an effect on tar decomposition. Comparing the catalytic effect, $Ca(OH)_2$ was superior to K_2CO_3. As a comparison, the effect on tar decomposition when $Ca(OH)_2$ was installed without changing the second reactor temperature was higher than that produced when the second reactor temperature was raised to 700–800°C. This result was also observed when the second reactor temperature was raised to 800–900°C and can be considered as the effect of $Ca(OH)_2$. In this study, $Ca(OH)_2$ exists in a state of CaO, such as high temperature, tar and secondary product char adhere to its surface [12]. Adhering material reacts with the pyrolysis product gas, thus gas is further produced. Additionally, the polarity of the active site of CaO can affect the π-electron cloud's stability of condensed aromatic compounds, and CaO itself has a cracking active site on both inner and outer surfaces [13].

4 Conclusions

In this study, the experimental apparatus connecting the two reaction tubes has been used for the study of heterogeneous reaction between tar and ash from cellulose pyrolysis and gasification modeling a fluidized bed gasifier. Alkali metal K and alkaline earth metal Ca were selected as the main ash contents that may act as the catalyst for tar decomposition. The catalytic effect of alkali metal K and alkaline earth metal Ca is that, not only they can change in the composition of the tar, but also reduce the condensable organic compounds. Additionally, K and Ca increase the amount of gaseous products. The catalytic effect of $Ca(OH)_2$ was comparatively superior to K_2CO_3. However, the role of K_2CO_3 might have been underestimated because K_2CO_3 reacts with silicon dioxide to form glass at temperature around 800°C. Based on the results of the present study, it can be concluded that K_2CO_3 and $Ca(OH)_2$ have the catalytic effect to decompose the tar and enhance gaseous production. In the case of using ash as a catalyst, it is necessary to take into account that the chlorides in the biomass are also included so that the change in the thermal decomposition properties is determined in further study.

Acknowledgements

Some of the work in this study was supported by the special funds for Basic Research (B) (No. 22404022, FY2010~2012) of Grant-in-Aid for Scientific Research of the Japanese Ministry of Education, Culture, Sports, Science and Technology (MEXT), Japan.

References

[1] Lin, S., World gasification process development for syngas production. *Journal of the Japan Institute of Energy*, **91**, pp. 826–834, 2012.

[2] Zhang, R., Brown, R.C., Suby, A. & Cummer, K., Catalytic destruction of tar in biomass derived producer gas. *Energy Conversion and Management*, **45**, pp. 995–1014, 2004.
[3] Han, J. & Kim, H., The reduction and control technology of tar during biomass gasification/pyrolysis: An overview. *Renewable and Sustainable Energy Reviews*, **12**, pp. 397–416, 2008.
[4] Hayashi, J., Iwatsuki, M., Morishita, K., Tsutsumi, A., Li, C. & Chiba, T., Roles of inherent metallic species in secondary reactions of tar and char during rapid pyrolysis of brown coals in a drop-tube reactor. *Fuel*, **81**, pp. 1977–1987, 2002.
[5] Shen, D.K. & Gu, S., The mechanism for thermal decomposition of cellulose and its main products. *Bioresource Technology*, **100**, pp. 6496–6504, 2009.
[6] Gomez-Barea, A. & Leckner, B., Modeling of biomass gasification in fluidized bed. *Progress in Energy and Combustion Science*, **36**, pp. 444–509, 2010.
[7] Yang, H., Yan, R., Chen, H., Lee, D.H. & Zheng, C., Characteristics of hemicelluloses, cellulose and lignin pyrolysis. *Fuel*, **86**, pp. 1781–1788, 2007.
[8] Hosoya, T., Kawamoto, H. & Saka, S., Pyrolysis gasification reactivities of primary tar and char fractions from cellulose and lignin as studied with a closed ampoule reactor. *Journal of Analytical and Applied Pyrolysis*, **83**, pp. 71–77, 2008.
[9] Levin, E.M., Robbins, C.R. & Mcmurdie, H.F., *Phase Diagrams for Ceramists, Volume I: Oxides and Salts*, American Ceramic Society, 1986.
[10] Bourhis, E.L., *Glass*, John Wiley & Sons, 2008.
[11] Lehman, R.L., Gentry, J.S. & Glumac, N.G., Thermal stability of potassium carbonate near its melting point. *Thermochimica Acta*, **316**, pp. 1–9, 1998.
[12] Widyawati, M., Church, T.L., Florin, N.H. & Harris, A.T., Hydrogen synthesis from biomass pyrolysis with in situ carbon dioxide capture using calcium oxide. *International Journal of Hydrogen Energy*, **36**, pp. 4800–4813, 2011.
[13] Tingyu, Z., Shouyu, Z., Jiejie, H. & Yang, W., Effect of calcium oxide on pyrolysis of coal in a fluidized bed. *Fuel Processing Technology*, **64**, pp. 271–284, 2000.

From biomass-rich residues into fuels and green chemicals via gasification and catalytic synthesis

S. C. Marie-Rose[1,2], E. Chornet[1,2], D. Lynch[2] & J.-M. Lavoie[1]
[1]*Université de Sherbrooke, Canada*
[2]*Enerkem Inc. Canada*

Abstract

Recycling carbon present in residual streams enhances sustainability and creates local wealth. Enerkem Inc. is a leading biomass gasification company headquartered in Montreal, Québec. The approach Enerkem has been developing involves: identification of low cost residual streams as feedstock, sorting, biotreatment (anaerobic and/or aerobic) and preparation of an ultimate residue (RDF). The latter is a rather uniform material that can be fed, as a fluff, to a bubbling bed gasifier in a staged gasification to carry out, sequentially, the needed chemical reactions that result in high syngas yields. The process can be adjusted to reach desired gas composition for synthesis or electricity generation as well as gas conditioning to produce an ultraclean syngas. Products for such a process are: (i) syngas with an appropriate range of H_2/CO ratios; (ii) CO_2 (which is recovered); (iii) solid char as a residue composed of the inorganic fraction of the raw material and some unconverted carbon that "coats" the inorganic matrices, and (iv) water that needs to be treated to meet the sewage specifications and thus be sent into the water distribution system of a given municipality. Enerkem is developing two parallel valorisation routes: (a) heat and power and (b) synthesis of (bio)methanol as a high yield product. The methanol is the platform intermediate that can be turned into ethanol (also with high yields), and other green chemicals. Yields of ethanol as the end product are above 350 liters/tonne of feedstock (dry basis) to the gasifier. Residual heat, also a product of the process, is used in the process itself and for district heating as well. The combined work of the Industrial Chair in Cellulosic Ethanol at the Université de Sherbrooke, focused on fundamentals in parallel and close relationship with Enerkem that led to technology development and implementation. The company has moved from bench scale, to pilot (150 kg/h in Sherbrooke, Québec), to demo (1500 kg/h in Westbury, Québec) to commercial implementation (12,500 kg/h in Edmonton, Alberta). The economics of the process are favourable at the above commercial capacity given the modular construction of the plant, reasonable operational costs and a tipping fee provided by the municipality for the conversion of the ultimate residue. When the project is in "production mode", Edmonton will have achieved 90% diversion of the urban waste from landfills and, furthermore the reduction in GHG (CO_2 equivalent) will be by 80% taking as reference the use of fossil fuel derived gasoline to yield the same energy output as the ethanol obtained by the Enerkem approach.

Keywords: alcohols synthesis, syngas production, methanol, ethanol, heterogeneous catalysis, gasification.

1 Introduction

Enerkem is a technology developer specialised in the gasification platform: feedstock preparation, feeding systems, syngas production, syngas clean-up, catalytic reforming and catalytic synthesis. Enerkem's technologies have been tested in the conversion of various waste products such as municipal solid waste, non-recyclable commingled plastics, residual biomass (from forest and agricultural operations) and numerous others carbonaceous residual streams to produce syngas, which is then used for the production of heat and/or power or for catalytic synthesis of biofuel (methanol and ethanol) and others specialty chemicals.

Biomass-derived ethanol plays an important role in reducing petroleum dependency and providing environmental benefits, through its role as fuel additive in the transportation fuel market. Ethanol is a winter fuel oxygenates and also an octane enhancer. For this reasons the Canadian government has legislated an objective of 5% of ethanol in gasoline for 2012. Aiming at this objective Enerkem's technology is an alternative to produce ethanol from non-food feed. At the same time is an alternative to landfill and incineration.

This paper focuses on characteristics of the syngas produced by gasification of RDF and on the catalytic synthesis of alcohols as second generation of biofuels.

2 Gasification of biomass

2.1 The basics

Gasification is the terminology used to describe the conversion of the organic matter present in waste and residues into a synthetic gas, a mixture of H_2, CO, CO_2, and low molecular weight hydrocarbons of formula C_xH_y. The gasifier is the vessel where the conversion is carried out.

The following chemical reactions are predominant during gasification:

1 Thermal decomposition (i.e. pyrolysis), which covers dehydration as well as cracking reactions leading to gases, intermediate vapours and carbon structures known as "char";

2 Partial oxidation of the "char" which forms CO and CO_2 generating heat for the otherwise endothermic reactions;

3 Steam-carbon, i.e. the water-gas reaction, that converts carbon structures into H_2 and CO;

4 Steam reforming of intermediates formed by thermal decomposition;

5 Reactions involving CO_2 and H_2 with carbon and with intermediates. Such reactions are kinetically slower than the steam induced reactions at the conditions used in gasifiers;

6 Water gas shift reactions that lead to a desired H_2/CO ratio.

The different gasification strategies and corresponding processes diverge on how to manage the six groups of chemical reactions stated above and how to generate and provide heat for the endothermic reactions.

Since thermal decomposition requires activation and, more important, the steam-carbon and steam-reforming reactions, as well as those with CO_2 are endothermic, heat has to be provided for gasification to proceed within reasonable reaction times (preferably seconds) to limit the size of vessels. Heat can be provided either indirectly (via steam, produced in a steam generation unit, or via the recirculation of a hot carrier, heated in a separate interconnected vessel) or directly (via air or oxygen injection). In the latter procedure heat is produced by oxidation of i) chemical species or functionalities present in the feedstock itself, ii) intermediates ("tar" and "char", the latter is also known as "pyrolytic carbon", formed by thermal decomposition, and/or iii) primary gaseous products. Since the amount of O_2 needed to satisfy thermal needs is below that needed for complete oxidation (i.e. combustion) of the feedstock, the net result can be considered, stoichiometrically, as a partial oxidation of the feedstock.

Reaction conditions: reactants concentrations, partial pressures, temperature, turbulence and reaction time are of paramount importance in defining the yields and product distribution from thermal gasification. The following "rules" are of essence:

1. When the process is conducted at low temperatures (<750°C) and short reaction times (of the order of seconds or even fractions of a second) thermal decomposition will predominantly produce oligomeric intermediates. The latter, will not be able, kinetically (low T and short t), to undergo subsequent secondary cracking or react with steam. Upon condensation of the oligomeric intermediates, the latter becomes the "primary tar". In such situation, the rate processes are those characteristic of low temperature fast pyrolysis which is known to produce large amounts of primary tar. The latter has a chemical structure derived from the constitutive macromolecules of the feedstock. In

primary tar from biomass, oligomeric anhydrosugars and lignin-derived oligomers are predominant.

2. At higher process temperatures (>750°C), the cracking of the intermediates produces high concentrations of free radicals whose recombination competes, kinetically, with steam (and CO_2) reforming reactions. When recombination reactions are predominant, the formation of large amounts of "secondary tar" is observed. Such secondary tar is of aromatic nature and it is often accompanied by soot, formed at high severities.

3. Uncatalyzed carbon-steam and carbon-CO_2 reactions are not kinetically significant, below 800°C. For this reason, activated carbon production, using steam, is carried out at >800°C. The presence of alkali catalyzes such reactions. This is important in biomass gasification since alkali (K being the most significant) is present in all biomasses. The point is that steam-carbon reactions can take place in a convenient temperature range when the reaction chemistry can be controlled by the presence of alkali. The choice of strong or weak alkali decides on the range and concentration (partial pressure) of steam to be used.

Reactor configuration and associated fluid-dynamics, which are characteristic of each gasifier type, are also of paramount importance in gasification since they are responsible for heat and mass transfer rates and define the residence time distribution patterns as well. Coupling effectively fluid-dynamics with reaction kinetics (which as previously discussed depends on temperature, pressure, concentration of reactants and reaction time) is the key to high yields and desired product selectivity.

2.2 The technology

The process commercialized by ENERKEM™ Inc., is the result of efforts started in the early 80s that led to the development of a core technology which couples fluid bed reactors with advanced gas conditioning strategies to provide a clean synthetic gas. The latter can be used for the co-generation of electricity and/or process heat. When combined cycles are used the electrical energy efficiency significantly exceeds that of the combustion/steam cycle. When the technology is coupled, in a staged-wise manner, with thermal and/or catalytic reforming it provides a mixture of H_2 + CO (syngas) that can be subsequently used for the catalytic synthesis of alcohols or hydrocarbons. Further water-gas shift leads to the production of H_2 upon removal of CO_2.

The technology can be applied to organic residues from diversified sources, such as sorted municipal solid waste (RDF), urban wood, agricultural residues, forest thinnings, sludges, as well as waste from various industries, such as sawdust and pulp mill residues, spent oils, plastic-rich residues and rubber-containing waste. The technology can also be applied to petroleum residues. It involves three main stages:

- adequate preparation of the raw material (size reduction and moisture adjustment, densification is optional);

- staged gasification initiated in a bubbling fluidized bed reactor, and pursued at increased severities in the freeboard and/or a secondary reforming unit;
- scrubbing or dry hot gas conditioning or a combination depending upon the end use.

2.3 How it works

Enerkem's gasification technology used low severity conditions to produce a crude syngas, followed by conditioning or cleaning of the crude syngas by subjecting the crude syngas to steam reforming, particulate removal, quenching, scrubbing, filtration and absorption prior to employing the syngas in the catalytic synthesis of alcohols (fig. 1).

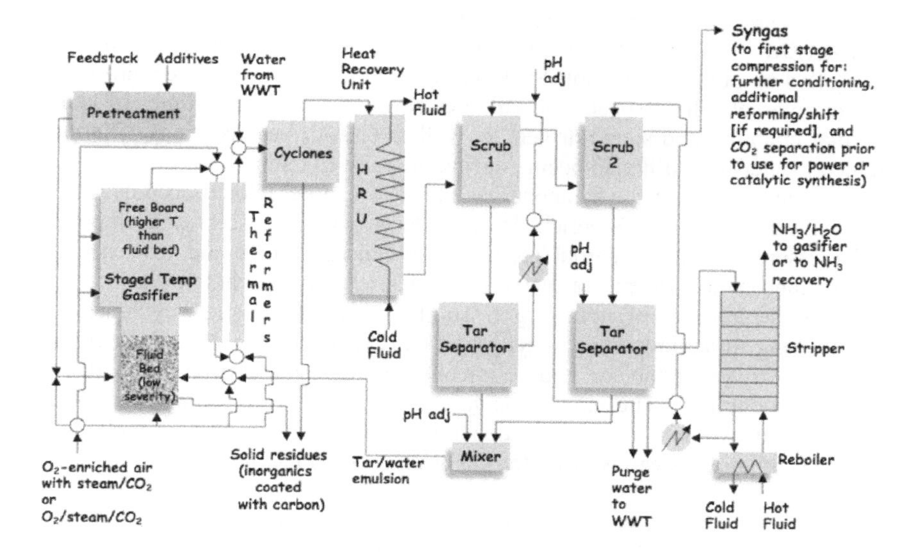

Figure 1: Syngas production from non-homogeneous biomass residues and waste.

The waste material has to be pretreated in order to obtain a feedstock with a characteristic particle size of about 5 cm as typical dimension. The process feedstock may need to be dried (using residual process heat) since its humidity at the reactor entrance should not exceed a specific level which is a function of the feedstock composition. The bulk density of the process feedstock needs to be typically higher than 0.15 kg/l for adequate feeding to the reactor.

The feedstock prepared as described above is directed towards the gasification reactor via an appropriately designed feeding system that controls the rate of material fed to a water-cooled transfer screw that injects the material into the fluid bed section where an appropriate fluidizing media is maintained.

Injection of the needed amounts of air or O_2-enriched air through a distributor grid located at the bottom of the fluid bed induces the fluidization patterns which result in high mixing and heat transfer rates which facilitate the reactions taking place during gasification. The quantity of air, O_2-enriched air, O_2-steam or O_2-steam-CO_2 required, depends on the organic composition of the residues. It is usually around 25% of the stoichiometric amount required for combustion of the organics. The temperature in the fluid bed part of the reactor is kept at about 700°C. The feedstock thermally decomposes producing volatiles (permanent gases, intermediates ranging from monomers to oligomers), and aerosols (entrained tar and small particles). The thermal decomposition also produces "pyrolytic carbon" (i.e. "char") that stays in the fluid bed until such time that its carbon content is decreased by the partial oxidation taken place in the bed and the particle terminal velocity reaches the level at which entrainment takes place towards the free board.

As the amount of carrier gas goes through the disengagement zone, in the ENERKEM™ process the temperature is increased by staged addition of controlled amounts of oxidant. Such addition continues in the freeboard or in a separate vessel. This exposes the volatiles and the entrained particles of char to thermal cracking and steam-driven, as well as CO_2-driven reactions. Thermally induced water-gas-shift also occurs during the raise in temperature.

The composition of the synthetic gas obtained can be tailored according to the gasification agent (air or O_2-enriched air), the feedstock composition, the temperature and pressure, the steam partial pressure, the fluid-dynamics and residence time distributions in the different zones of the reactor as shown in table 1.

After separation of the larger solid particulates using cyclones, gas conditioning to achieve a clean syngas is carried out via a two-stage scrubbing process, recovering the tar and re-injecting it into the gasifier for additional syngas production. The clean syngas will contain low molecular weight hydrocarbons whose concentration depends on the nature of the feedstock.

The synthetic gas can be used for the production of energy as follows:

- combusted as is, or in combination with natural gas, using commercial boilers to produce process heat (steam);
- as a fuel, alone or with natural gas, in internal combustion engines (ICE) to produce electrical energy. Heat from the ICE hot exhaust gases can be recovered and used for steam (or other fluid) which leads to combined cycle or co-generation applications;
- as a fuel, alone (if the calorific value is sufficiently high, as when O_2-enriched air or O_2-steam is used rather than normal air) or with natural gas, in gas turbines to produce electrical energy via combined cycles or in co-generation mode;
- as a feedstock (when O_2-steam or O_2-steam-CO_2 are used) for the synthesis of alcohols or hydrocarbons. For this purpose the low molecular weight hydrocarbons need to be reformed.

Table 1: Biomass and syngas composition, in low severity gasification.

Typical biomass composition	wt%
Volatile matter	>70
Ash	1.5
Moisture	20
Ultimate analysis	**wt%**
C	54.2
H	6
O	38.9
N	0.3
S	0.1
Gasification conditions	
O_2/Steam/CO_2 gasification, low severity+ thermal cracking and reforming at higher severity	
Syngas composition	**mol%**
N_2	2 - 6
Ar/O_2	<1
H_2	23 – 27
CO	21 – 23
CO_2	38 – 44
CH_4	6 – 8
C_2H_4	Traces
C2 – C5	0.2 – 0.5
C5 – C10	Traces
Input C converted into syngas	>90%

3 Syngas conversion into biofuel

Conversion of syngas to liquid fuels as well as conversion rates are directly related to the composition of the catalyst. Syngas can be efficiently converted to different products as alcohols, DME and hydrocarbons (fig. 2). Although several routes are available, a promising route at the industrial level is the production of methanol since it is a selective conversion with established catalysts and proceeds at productivity levels higher than 1 kg methanol per kg of catalyst per hour.

3.1 Syngas to methanol

The conversion of synthesis gas to methanol is thermodynamically constrained hence the need to recycle the unconverted gas to achieve high CO conversion

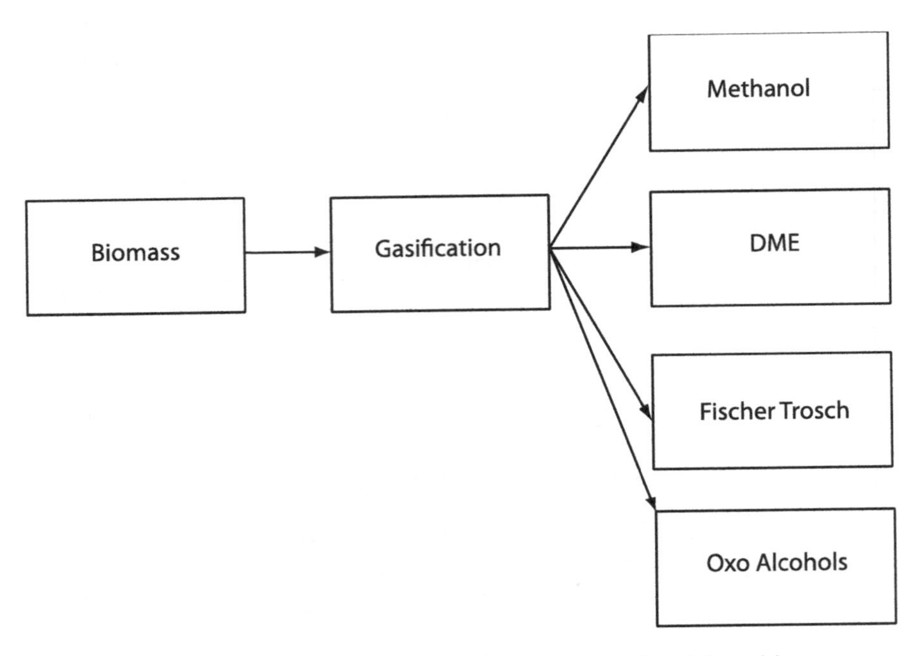

Figure 2: Examples of applications for syngas produced from biomass.

levels. This translates into high reactor volume and increased compression costs. Methanol synthesis was carried out with the conditioned syngas using a three-phase reactor (slurry bubble column) in which a $Cu/ZnO/Al_2O_3$ catalyst was suspended. Before the synthesis of methanol, the catalyst is reduced in hydrogen. The reductive and active form of the catalyst is $CuO/ZnO/Al_2O_3$. The principal stoichiometric reactions involved in this chemical conversion are:

$$CO + 2H_2 \leftrightarrow CH_3OH \quad \Delta H_{298K} = -91 \text{ kJ mol}^{-1}$$

It is subjected to a thermodynamic equilibrium that limits the process to low conversion per pass (CO conversion about 45%) and therefore, implies a large recycle of unconverted gas. The reaction is strongly exothermic and consequently, requires significant cooling duty.

4 Methanol to ethanol

4.1 Methyl acetate as intermediate of ethanol

Carbonylation of methanol to acetic acid or methyl acetate is a well known industrial process. It is currently performed, industrially, in a homogeneous reaction system where a catalyst is dissolved or suspended. The Monsanto/Celanese/Eastman and BP Chemicals Cativa™ processes in the presence of rhodium and iridium respectively are conducted at moderate temperature and pressure in the presence of halide (iodide) promoters [1]. The mechanism of the reaction of acetic acid synthesis is well described in the literature and it is known as Monsanto cycle [2]. The formation of methyl acetate can be explained by the reaction between the acyl iodide formed during the reaction and the excess of methanol present in the feed.

Heterogeneous systems for methanol carbonylation have been suggested by investigators for several years. Research has been primarily focused on two possible catalysts for this reaction: rhodium supported on polymers (Acetica process is based on this principle) or zeolites and a variety of metals supported on activated carbon [3]. The choice of support seems to play an important role in the activity of the catalyst [4, 5].

In the "Enerkem process" [6] carbonylation of methanol is carried out maintaining methyl alcohol in vapour phase using a fixed bed packed with a rhodium-based catalyst. The methanol is vaporized under pressure and mixed with the CO-rich fraction prior to flowing through the reactor. The methanol to CO molar ratio is comprised between 1 and 4, whereas methyl iodide (co-catalyst) added to methanol is maintained at a molar ratio between 1 and 5 wt% relative to the methanol. The operating conditions are such that the GHSV (based on CO) varies between 100 to 2,000 h^{-1}, at temperatures comprised between 170 to 300°C and total pressures from 10 to 50 atm. It was reported that the CO is converted at a rate near 100% when the methanol/CO ratio is >2. The selectivity varies as a function of temperature and pressure. It was found that within a wide range (200–240°C, 15–50 atm) for the specified GHSV (Based on CO) range, there was molar selectivity of 50–75% methyl acetate and 25–50% of acetic acid.

4.2 Methyl acetate hydrogenolysis: synthesis of ethanol

Methyl acetate produced as previously via carbonylation, is maintained in liquid form at 20°C. It is pumped in a pressure vessel ranging from 10 to 50 atm, through a heat exchanger that vaporizes it at temperatures ranging from 150 to 225°C. Preheated hydrogen in the same temperature range is mixed with the methyl acetate vapour at the exit of the heat exchanger. The molar ratio H_2 to methyl acetate is from 5 to 11. The hot mixture is flown through a catalytic bed where a CuO/Cr_2O_3 or $CuO/ZnO/Al_2O_3$ catalyst is placed. The CuO is reduced with H_2/N_2 mixtures prior to adding any methyl acetate [7]. Methyl acetate is converted to methanol and to ethanol at GHSV (based on H_2) comprised between 1,000 and 2,000 h^{-1} and the methyl acetate conversion is up to 95%, with a higher selectivity in ethanol.

5 Conclusion

Enerkem has successfully proven, at pilot scale, the feasibility of converting heterogeneous biomass residues into a homogeneous syngas that can be used for the synthesis of methanol. The latter is a key intermediate building block for chemical commodities and biofuels such as bio-methanol and from the latter bio-ethanol. The technology is applicable to any organic material and residual biomass including urban biomass such as residential waste. It is preferable that the waste sorted to remove recyclables as well as ferrous metals, glass and ceramics. Typically, to maintain consistent performance, inorganic matter levels should be kept below 20 wt% in the feedstock (dry basis).

References

[1] Ormsby, G., Hargreaves, J.S.J. & Ditzel, E.J., A methanol-only route to acetic acid. *Catalysis Communications*, **10**, pp. 1292–1295, 2009.

[2] Kollár, L., *Modern Carbonylation Methods*, Wiley-VCH, 2008.

[3] Merenov, A.S. & Abraham, M.A., Catalyzing the carbonylation of methanol using a heterogeneous vapor phase catalyst. *Catalysis Today*, **40**, pp. 397–404, 1998.

[4] Fahey, D.R., ed., *Industrial Chemicals via C1 Processes*, American Chemical Society, 1987.

[5] Yashima, T., Orikasa, Y., Takahashi, N. & Hara, N., Vapor phase carbonylation of methanol over RhY zeolite. *Journal of Catalysis*, **59**, pp. 53–60, 1979.

[6] Chornet, E., Valsecchi, B., Avilla, Y., Nguyen, B. & Lavoie, J.-M., Production of ethanol from methanol. US Patent Application US 2009/0326080, 2009.

[7] Claus, P., Lucas, M., Lücke, B., Berndt, T. & Birke, P., Selective hydrogenolysis of methyl and ethyl acetate in the gas phase on copper and supported Group VIII metal catalysts. *Applied Catalysis A: General*, **79**, pp. 1–18, 1991.

Thermal gasification of agro-industrial residues

P. S. D. Brito, L. F. Rodrigues, L. Calado & A. S. Oliveira
Polytechnic Institute of Portalegre, Portugal

Abstract

Farming and the agro-industrial sector generate large amounts of organic residues which have a high energetic potential and in most cases are not recovered. Various technologies, including those for energy and agricultural applications, have been considered for the economical valorization of these residues. This work contains the results of a study on the potential use of some agro-industrial residues produced in the Portuguese Alto-Alentejo region, in particular, coffee husks as well as forest and vineyards residues. The study was conducted in a pilot thermal gasification plant, installed at Portalegre's Industrial Park, based on the fluidized bed technology, with a processing capacity of 70 kg/h, and operating at around 800°C. The gasification tests were performed continuously for several days, using different wastes in order to optimize the heat value and composition of produced syngas. The economic viability of the use of these wastes as row material for the commercial production of synthesis gas and synthetic city gas was comparatively evaluated. The results achieved allow the conclusion that the waste biomasses studied show interesting features for use as syngas row-material.
Keywords: biofuel, pilot plant, gasification, fluidized bed, syngas.

1 Introduction

Bioenergy and biofuels are one of the most important alternatives to fuel production to minimize the continuous depletion of fossil fuels resources and the increasing concerns with environmental issues. Organic living matter, or biomass, in energy terms, is no more than a form of chemical storage of solar electromagnetic energy. When the bonds between carbon, hydrogen and oxygen atoms of the organic molecules are broken, during technological processes such as digestion, combustion or gasification, there is a release of the stored chemical energy. It is estimated that the annual global biomass energy storage is approximately 1.33×10^{14} W (0.26 W/m^2), 57.14% of which is produced on earth. It is estimated that the biomass accumulated as energy is of the order of 1.5×10^{22} J. It is obvious that only a part of this biomass can be exploited in technological and economic terms for the production of energy. However, there is a large amount of biomass available that possesses excellent conditions of exploitation, particularly, wastes from agriculture, forestry and related industries, as well as, from industries and domestic wastes in general. According to the European Environmental Agency (EEA), the use of biomass for energy purposes will grow significantly over the coming decades. Predictions point, for the end of the century, to a share of about 25% for bioenergies among all energy sources, namely, oil, natural gas, renewable and nuclear.

Gasification appears to be a promising process to convert biomass to syngas containing methane and hydrogen to be used directly as energy source or as row material for the production of liquid fuels and other chemicals [1, 2]. The production of gas with calorific value from materials containing carbon is a quite old technology that is known since the end of the 19th century. With the development of technologies that use oil as row material, the interest in biomass gasification processes declined until the energy crisis, in the 70 years of the last century. More recently, due to the discussion on climate changes, the Kyoto Protocol commitments and the need to use renewable energy sources, the interest in gasification of biomass is increasing rapidly [1, 3]. Biomass gasification offers advantages over combustion processes since that technology allows the coproduction of minor levels of NOx, SOx, particulates and heavy metals in gaseous emissions and presents a greater potential for reducing CO_2 emissions.

Different reactor types have been used for carrying out biomass gasification: fixed bed (updraft, downdraft, or cross-current), fluidized bed, doubled fluidized bed, circulating fluidized bed, entrained flow, among others [1, 3]. The use of fluidized bed technology in biomass gasification units makes it possible to use relatively smaller gasifiers and larger capacities than with fixed bed. The higher efficiency of fluidized bed comes from its higher specific reaction area when compared to fixed bed. Fluidized beds also proved to permit more versatile units in terms of feeding materials of different characteristics and origins, although the relatively high temperatures used to avoid sintering problems of the bed material leads to the formation of tars that complicate operation of equipment due to clogging of equipment and piping [4].

Agroindustry is one of the most important economic activities in the Portugal's Alentejo region; it generates large amounts of residues, in particular, vine prunings, bagasses, coffee husks, forest residues, among others, which require treatment or an adequate recovery to minimize environment impacts and increase the economic value of these wastes. A large variety of technologies has been developed over the past decades to deal with this problem. Among the proposed technologies, those oriented towards energy recovery, including combustion and gasification of biomasses has attracted much interest [5].

The purpose of this work is to contribute to an assessment of the potential of biomass energy available in the Norte Alentejo region and to study the technical feasibility of energetic recovery of wastes produced in agroindustry through thermal gasification technologies.

2 Experiments

The experiments were performed in a gasification pilot plant (fig. 1), which is based on an up-flow fluidized bed gasifier, operated up to 850°C, under a total pressure below 1 bar and at a maximum pellet feeding rate of 70 kg/h. In the present work, the results of gasification experiments of different compositions of mixtures of vine pruning and wood pellets are presented.

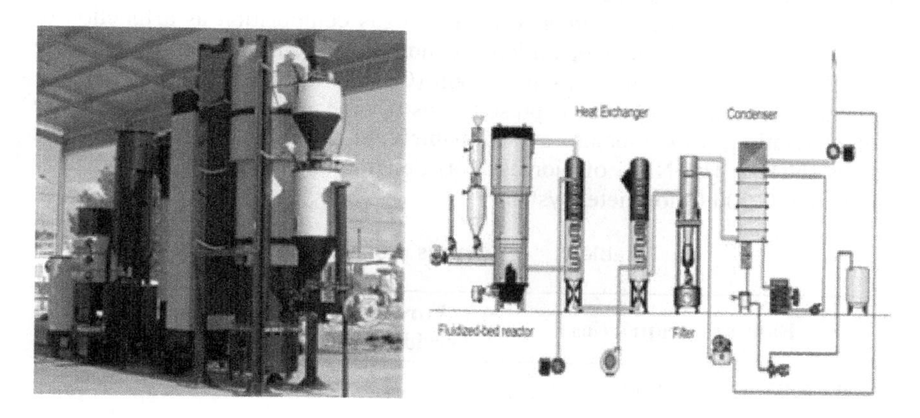

Figure 1: Biomass gasification pilot scale plant at Escola Superior de Tecnologia e Gestão, Instituto Politécnico de Portalegre, Portalegre, Portugal. The main components of the unit are described in the text.

Figure 1 displays a diagram of the biomass gasification pilot scale plant used in the experiments. The main components of the unit are the following: a) *Biomass feeding system* with two in-line storage tanks that allow the discharge of the biomass into the reactor using an Arquimedes screw at a variable and controllable speed; these two storage tanks act as buffers to avoid the entrance of air through the feeding system; b) *Fluidized bed reactor* that is a tubular reactor 0.4 m in diameter and 2.5 m height, internally coated with ceramic refractory materials; biomass enters the reactor at the height of 0.5 m, from its base, and preheated air enters the reactor coming from the base through a set of diffusers, warranting a flow of about 70 m^3/h; three temperature sensors are installed inside the reactor in order to monitor and control the gasification temperature around 800°C; syngas leaves the reactor trough his top, at about 700°C; as referred, the reactor operates at negative pressure gradient produced by a vacuum pump installed at the end of the process line; temperature control inside the reactor is a determining factor for the gasification process and this control is achieved by tuning the amount of air admitted to the reactor, namely, increasing the air flow rate to increase the reaction temperature; the fluidized bed was made of 70 kg of dolomite. c) *Gas cooling system* consisting of two heat exchangers; the first exchanger cools the syngas to about 300°C using an air co-current flow that enters the unit, and the second heat exchanger cools the syngas down to 150°C by forced flow of air coming from the exterior; d) Cellulosic *bag filter* that allows the removal of carbon black and ashes particles produced during the gasification process; the cleaning of the filter is made by pressurized syngas injection; black carbon is collected at the bottom of the bag filter and is stored in a proper tank; e) *Condenser* where liquid condensates are removed by cooling the syngas down to room temperature on a third tube heat exchanger.

Tests were made using feeds of different biomasses such as coffee husk, forest residues and vineyards residues, at 800°C. Feedstock admission rates of 40 and 63 kg/h were tested in order to study syngas composition as a function of feedstock composition and operational conditions. Biomasses tested have the elemental composition, in terms of carbon (C), hydrogen (H) and oxygen (O), presented in table 1. Table 1 presents also some other parameters for the characterization of the biomasses used, namely, humidity, density and Net Heat Values (NHV). The NHV of biomass was determined with an IKA Laboratory Equipment C 200 Calorimeter system.

Table 1: Biomass proprieties.

Biomass proprieties	**Forest residues**	**Coffee husk**	**Vines pruning**
Elementary analysis (%)			
N	2.4	5.2	2.6
C	43.0	40.1	41.3
H	5.0	5.6	5.5
O	49.6	49.1	50.6
Humidity (%)	11.3	25.3	13.3
Density (Kg/m^3)	650	500	265
Net Heat Value (MJ/Kg biomass)	21.2	20.9	15.1

Syngas analysis was performed in a Varian 450-GC gas chromatograph with two TCD detectors that allow the detection of H_2, CO, CO_2, CH_4, O_2, N_2, C_2H_6, C_2H_4 (equipped respectively with CP81069, CP81071, CP81072, CP81073 and CP81025 Varian GC columns), using helium as carrier gas. Syngas samples for the analysis referred to above were collected in appropriate collection and analysis *Tedlar*© bags at the condenser exit every time when the gasification of a given feedstock composition reached its stationary state. Collected syngas samples were injected (within 1 hours after sampling) directly from the sampling bags in the chromatograph using a peristaltic pump operating at is maximum rate and equipped with a *Marpren*® tube. Chromatographic peaks for the different gases under analysis were identified based on their retention times, and by comparing them with the retention times of the same gases in the reference chromatogram of the "custom solution", provided by Varian. Gas mass percentage composition was calculated on the basis of peak areas under chromatographic signal.

3 Results and discussion

3.1 Gasification

Compositions and net heating value (NHV) of the different syngas produced are presented in table 2. NHV's were calculated on the basis of syngas composition determined by gas chromatography and using the standard combustion heats of the compounds obtained, at 25°C.

Generally, the results show that the produced syngas is relatively rich in carbon monoxide (mass percentage between 8 to 19%), methane (1 to 5%) and hydrogen (5 to 13%), and therefore these three gases are the main responsible for the heat content of the syngas produced. NHV's values lie between 2 and 5 MJ/kg, as can be seen in table 2. On the other hand, it can be seen that the syngas contains also large amounts of nitrogen (48 to 56%) and some carbon dioxide (16 to 20%) resulting from the partial combustion process that takes place simultaneously with biomass gasification.

The thermal biomass gasification process involves a set of complex chemical reactions that lead to the formation of three fractions: the syngas, ashes (chars) and condensates. The most important fraction, amounting to more than 70% (wt) is made of light gases, namely, CO, H_2, CH_4, CO_2, and N_2. In fact, it is considered that in the thermal gasification process, gases leave the reactor in equilibrium condition. Main equilibrium reactions to be considered are the steam methane reforming reaction (1) and primary water-gas reaction (2):

$$CH_4 + H_2O \leftrightarrow 3H_2 + CO \quad (1)$$

$$CO + H_2O \leftrightarrow H_2 + CO_2 \quad (2)$$

In a first approach, it may be said that the gas composition, the syngas fraction and the amount of aches and condensates are a function of biomass nature, its elemental composition and of the gasification operational conditions, in particular, of the process temperature.

Generically, the results obtained with the various biomass studied show a relation between the biomass calorific value and the calorific value of the syngas obtained: higher biomass calorific value results in higher calorific syngas production. In fact, it turns out that the syngas formed with the highest calorific value was obtained by gasification of forest residues, a more energetic material. This relationship between the biomass calorific content and the syngas NHV can be explained considering first, that, the biomass calorific value is related to the

Table 2: Experimental conditions and syngas analyses.

Experimental conditions	Forest residues			Coffee husk			Vines pruning		
Temperature (ºC)	815	815	790	815	790	790	790	790	815
Admission Biomass (Kg/h)	63	74	63	28	28	41	25	55	55
Air Flow Rate (Nm^3/h)	94	98	98	75	72	80	52	40	40
Ratio $O_2/O_{2\ Stoichiometric}$	1.11	0.99	1.16	2.63	2.52	1.96	2.96	0.58	0.58
Syngas flow rate (Nm^3/h)	106	94	100	106	88	116	107	108	102
Syngas fraction (dry basis)									
H_2	8.2	8.4	7.6	12.4	7.6	7.5	5.1	10.4	12.7
CO	18.6	18.0	17.9	11.4	11.1	10.6	8.3	11.7	14.1
CH_4	4.6	4.4	4.4	1.6	2.4	2.4	1.1	2.4	2.3
CO_2	16.7	17.1	17.1	18.7	17.0	18.5	16.5	20.1	17.9
N_2	48.0	48.2	49.2	52.3	54.2	55.2	56.4	51.2	49.1
Syngas composition (g/kg dry biomass)									
CO	438	323	399	711	572	506	855	328	375
H_2	14	11	12	55	28	25	38	21	24
CH_4	62	45	56	56	71	65	63	39	36
CO_2	620	483	602	1836	1380	1390	2675	888	751
Syngas NHV									
MJ/Nm^3	5.16	5.02	4.93	3.34	3.20	3.07	1.99	3.46	4.02
KWh/Kg	1.13	1.09	1.07	0.73	0.65	0.63	0.37	0.73	0.89
Cold Gasification Efficiency	0.41	0.30	0.37	0.60	0.47	0.42	0.95	0.45	0.49

amount of carbon and hydrogen present in the biomass, i.e. an increased amount of carbon and hydrogen, lead to higher calorific value, and, secondly, a larger amount of these two elements allows production of larger quantities of hydrogen and carbon monoxide, the major contributors for the calorific value of the syngas.

Analysis of the dependence of producer gas properties on conditions of biomass admission shows that for all biomasses studied there is a decrease in syngas calorific value with increasing quantity of biomass admitted. This is consistent with the fact that a faster acceptance of biomass by the reactor reduces the extent of reaction and the equilibriums represented by equations (1) and (2) are not attained.

For forest residues and coffee husks, an increase in gasification temperature promoted the formation of a syngas with higher hydrogen and carbon monoxide contents and, consequently, higher NHV of the syngas. This could be a result of a shifting in the gasification equilibrium, since reaction represented by equation (2) is endoenergetic and, therefore favoured by a temperature increase. For these residues cold gasification efficiency values are greater than 0.3 and increase with increasing gasification temperature and decreasing oxygen to biomass ratio.

Results of the study of vine pruning gasification (table 2) shows a decrease in the NHV when the gasification temperature is increased. It should be noted that the decreases in NHV were accompanied by decreases in carbon monoxide production and consequent increase in carbon dioxide content in the producer syngas.

3.2 Economic evaluation

In this section the achieved results are analyzed from an economic perspective in a framework of a demand for alternative energy sources to fossil fuels in order to minimize environmental problems and make the full use of the potential of this technology. The analysis will focus on Portuguese Alto-Alentejo main industries, in particular, agro-industries, which are in essence small companies with a high territorial dispersion. In this approach, the search for alternative energy sources must imply a strategy to develop local energy sources, in particular, from industrial waste, in order to obtain a decentralized generation of electricity, liquid or gaseous fuels. For analysis purposes, it will be considered that the aim would be the use of thermal gasification to produce a gaseous fuel which will substitute natural gas utilized by many regional industries.

The analysis was based on the results obtained in this work on gasification of different biomasses, in particular, on the composition and calorific value of syngas obtained and on present prices of natural gas at market place. A daily production of 1 ton of waste was assumed. Table 3 shows the parameters used in the calculation and the results in terms of annual return for each of the biomasses.

It was assumed that a gasification plant with a capacity of 1 ton per day has an investment cost of around 80 k€, constant prices, and that the initial investment would be amortized over about 5–6 years, the average value for other renewable energies systems, such as photovoltaic and micro-wind energy.

Table 3: Economic evaluations for biomass gasification unit.

Parameters			
Biomass (ton/day)		1	
Price (€/KWh)		0.0501	
Fixed costs (€/month)		10	
Energy outflow in operation (%)		15	
Biomass	**Forest residues**	**Coffee husk**	**Vines pruning**
Advantage (k€/year)	17.6	17.1	16.7

4 Conclusions

Based on the results achieved in this work it may be concluded that:

- The increase of gasification temperature and feedstock admission rate improved the syngas Net Heat Value (NHV);
- the economic evaluation for the recovery of agroindustrial wastes by thermal gasification compared with the present costs for natural gas in the Alto-Alentejo region industries shows clear economic potentialities for the application of the biomass gasification technology considered.

Acknowledgements

The authors give thanks to the U.E. Commission, POCTEC Program, ALTERCEXA Project, for financial support.

References

[1] McKendry, P., Energy production from biomass (part 3): gasification technologies. *Bioresource Technology*, **83**, pp. 55–63, 2002.

[2] Rajvanshi, A.K., Biomass gasification. *Alternative Energy in Agriculture*, Vol. II, ed. D.Y. Goswami, CRC Press, pp. 83–102, 1986.

[3] Balat, M., Gasification of biomass to produce gaseous products. *Energy Sources, Part A*, **31(6)**, pp. 516–526, 2009.

[4] Prins, M.J., Thermodynamic analysis of biomass gasification and torrefaction, PhD Thesis, Technische Universiteit Eindhoven, Proefschrift, 2005.

[5] Fiori, L. & Florio, L., Gasification and combustion of grape marc: comparison among different scenarios. *Waste and Biomass Valorization*, **1(2)**, pp. 191–200, 2010.

Process analysis of waste bamboo materials using solvent liquefaction

Q. Wang[1], Q. Qiao[1], Q. Chen[1], N. Mitsumura[1], H. Kurokawa[1], K. Sekiguchi[1] & K. Sugiyama[2]
[1]*Graduate School of Science and Engineering, Saitama University, Japan*
[2]*Hachinohe National College of Technology, Japan*

Abstract

Bamboo is one of the significant biomass resources; it has been used in houses, flooring, construction of scaffolding and bridges, among others. The solvent liquefaction process is one of the promising techniques for effective utilization of waste bamboo materials for the lignocelluloses which can be converted to liquid reactive materials as biomass-based materials. Bamboo has the advantage of providing the liquefied products with a small range of variances. The components of bamboo have high acidity in the presence of mineral acid catalysts and possess the constituents which can react with polyethylene glycol 400 (PEG 400). In this study, waste bamboo materials have been used in liquefaction experiments. The liquefaction process and liquefied residue have been monitored according to the liquefied conditions and surface changes of waste bamboo samples observed by a scanning electron microscope. The changes in the functional groups have been analysed by a Fourier transform infrared spectrometer and behavior of the crystalline structures of liquefied bamboo has been determined by X-ray diffraction. Other experiments, such as degree of polymerization have also been carried out for confirming the results. Regarding the results, it was found that an increment of the temperature and the amount of the acid catalysts improved the efficiency of liquefaction. At the same time, the dissolution time of lignin was significantly shorter than the one of cellulose in the solvent liquefaction process of PEG 400.
Keywords: bamboo, liquefaction, biomass-based materials, FT-IR, X-ray diffraction, SEM.

1 Introduction

Bamboo, as a significant biomass resource, has been used in many fields of biomass utilization. It is also extensively used to make furniture, food steamers, chopsticks, paper pulp. Many kinds of bamboo grow naturally in tropical, subtropical, and temperate regions around the world. Bamboo is fast-growing, economic and of high cultural significance in East Asia and South-Eastern Asia. The solvent liquefaction process is one of the promising techniques for effective utilization of bamboo for the lignocelluloses can be converted to liquid reactive materials as biomass-based materials. Liquefaction technology, as one of the effective methods of using the lignocellulosic biomass, was mentioned for preparation biomass-based materials. In early studies of organic solvents liquefaction, Lin *et al.* [1] had also attempted to explain the reaction liquefaction mechanism of cellulose with phenol solvent under the acid-catalyzed conditions. It was found that the yield of various compounds products was dependent on the reaction conditions. Therefore, it may control the structure and properties of liquefied products in the later stage of liquefaction by controlling these reaction conditions or time. As target products, biomass-base polyurethane films and Novolak phenol foam resins were prepared, their structures and properties were discussed [2, 3]. The research of bamboo liquefaction in this area is still relatively small. Moreover, it is known that bamboo fiber has high strength, low elongation, and high crystalline. And the high crystallinity leads the solvent liquefaction to develop slowly and inefficiently. A longer time is needed to liquefy higher crystalline index biomass such as bamboo materials [4]. Therefore, it is necessary to understand the change of waste bamboo materials in the liquefaction process in order to promote the liquefaction process and efficiency. Nevertheless, the mechanism of forming condensed residues still needs be clarified [5] and is still driving the research to overcome the problems arising from drawbacks in the physical and mechanical properties of resin products from liquefied biomass [6].

In this study, waste bamboo materials were used in liquefaction experiments. The liquefaction process and liquefied residue measured according to the conditions and surface change of waste bamboo materials have been observed by a scanning electron microscope (SEM). The changes in the functional groups were analyzed by a Fourier transform infrared (FT-IR) spectrometer and influences on the crystalline structures of liquefied bamboo were determined by the X-ray diffraction (XRD).

2 Materials and methods

2.1 Materials

Waste bamboo materials such as used bamboo chopsticks had been collected from the waste treatment and recycle factories of Japan. Experimental samples of cellulose (CAS No: 9004-34-6 MP Blomedicais, LLC, Co, Ltd., Japan) and lignin (Alkaline, CAS No: 9005-53-2 TCI. Co, Ltd., Japan) were used as the simulated bamboo material. Oven-dried waste bamboo powder, milled in smashing equipment and retained at the size ranges below 0.5 mm, was used as sample in experiments. All the samples of waste bamboo powder were dried in an oven at 105°C for 24 hours. The main different acidic catalysts such as sulfuric acid (95 wt.%), PEG 400 and other reagents were prepared with the analytical grade reagents made by Wako Pure Chemicals, Co. Ltd., Japan in accordance with the Japanese Industrial Standard (JIS).

2.2 Chemical composition of waste bamboo materials

In the proximate and ultimate analysis, the analysis of ash content of the residue left after the combustion of bamboo sample was performed according to the industrial standard method (JIS-M8812). 14.4 wt.% of moisture content in bamboo sample was determined by measuring the weight loss after drying the sample at 105°C. 3.8 wt.% of ash was measured by heating the bamboo powder to 900°C under carefully controlled conditions. 29.6 wt.% of 1% NaOH soluble composition and 11.9 wt.% of methanol-benzene soluble was determined in this study. As components of bamboo, 68.9 wt.% of holocellulose and 18.9 wt.% of lignin were present.

2.3 Liquefaction experimental method

The samples of waste bamboo powder were used in liquefaction experiment and the weight charge ratio of bamboo and PEG 400 was 1/4 and the reaction temperatures were 120°C and 150°C by oil bath. The raw materials put into a 500 ml three necked flask equipped with a stirrer. Reactants were treated with methanol as the diluents. The different amount of catalyst from 3 wt.% to 10 wt.% was added to the reactant as the reaction start time, and added the methanol into liquefied products as the stop of reaction time.

2.4 Measurement of liquefaction residue

The liquefied bamboo samples were weighted and diluted with methanol, and Whatman filter papers (No. 40) were used to collect the liquefied residue. The liquefied residue was dried by heating in an oven at 105°C in 12 hours and calculated using the equation

$$\text{Liquefied ratio (wt.\%)} = (\text{Solvent-insoluble residue} / \text{Raw material}) \times 100 \quad (1)$$

where the solvent-insoluble residue is the weight of liquefied residue through filtration which is diluted by methanol from liquefied products, and the raw material is the weight of the reactants.

2.5 Determination of the degree of polymerization

About 0.04 g liquefied bamboo at different experiment conditions were dissolved in 10 ml cupriethylenediamine hydroxide solution. Then, the intrinsic viscosity of the solution was measured three times by a handmade Ubbelodhe viscometer, and the degree of polymerization (DP) was calculated according to the flow time. All the measurement experiments were more than 3 times for each average value.

2.6 Fourier transform infrared spectrometer analysis

Several liquefied bamboo samples were analyzed using a Fourier transform infrared (FT-IR) spectrometer (FT-IR-6100 Jasco, Co. Ltd., Japan). The ratio of liquefied bamboo samples and spectroscopic grade KBr was 1:100; the time of mixing sample was more than one hour for full mixed. All of the infrared spectra were recorded in absorbance units in the 4000– 400 cm^{-1} range.

2.7 Scanning electron microscope analysis

The surface changes of native and liquefied bamboo was observed by Hitachi (SEM, Model S-2400, Hitachi Co. Ltd., Japan) scanning electron microscopes (SEM) at an acceleration voltage of 15 kV.

2.8 X-ray diffraction analysis

Crystalline structures of bamboo and liquefied bamboo were analyzed by Ultima III X-Ray Diffractometer (Rigaku Co. Ltd., Japan). Ni-filtered Cu Kα radiation (λ = 0.1542 nm) generated at a voltage of 40 kV and a current of 40 mA was used. Intensities in the range from 10° to 40° with 4°/min scan speed was set of total X-ray diffraction (XRD) analysis experiment. The crystallite height 200 (h_{200}) and amorphous height (h_{am}) was used to calculate the crystalline index (CI) from the equation

$$\text{Crystalline index (CI) (\%)} = \left(\frac{h_{200} - h_{am}}{h_{200}} \right) \times 100 \quad (2)$$

3 Results and discussion

3.1 Residues results in different experimental conditions

The amount of liquefied residue (LR) was a signification index to evaluate the performance of liquefied bamboo in solvent liquefaction. Two variables were used in the experiment. One was different amount of sulfuric acid, another was different reaction temperature. Two response variables had impact on the liquefaction results. Figure 1 shows the liquefaction rate of bamboo related to different amount of sulfuric acid within 90 min at 120°C. A greater quantity of sulfuric acid was added to reactant, but no fewer residues have been obtained. On the contrary, the residues increased from 30 min. The cellulose gradually converted to levulinic acid derivatives, following the conversion of lignin into solvent-soluble compounds in the early stage of bamboo liquefaction. Then, the mixture of levulinic acid derivatives and aromatic compounds reacted and converted to the solvent-soluble residue. The result indicated that the insoluble residue was formed when cellulose and lignin coexisted [7]. A condensation reaction had occurred during the liquefaction process if the experimental conditions, such as solvent/cellulose addition ratio, were changed [8].

When 5 wt.% sulfuric acid of solvent was added into the experiment, more residues were obtained; this also indicated that the insoluble residue was contained in the liquefied products. This problem will affect the mechanical properties of resin products from liquefied biomass [6].

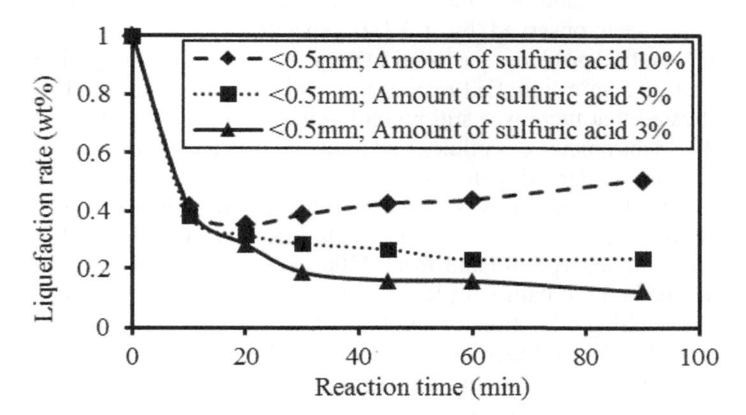

Figure 1: Liquefaction rate of waste bamboo materials in different amount of sulfuric acid within 90 min at 120°C.

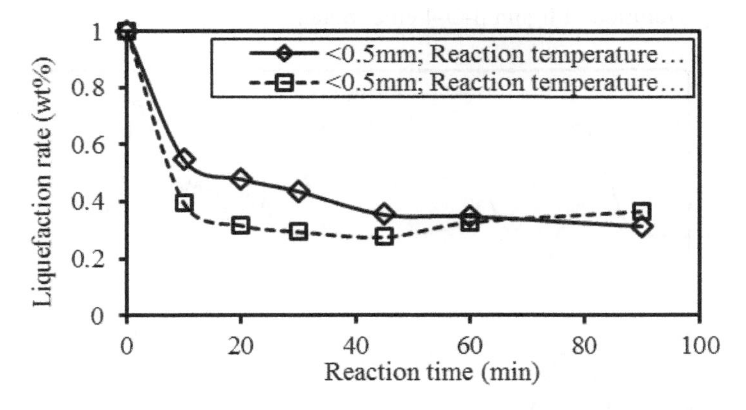

Figure 2: Liquefaction rate of waste bamboo materials with different reaction temperature within 90 min and 5 wt.% of catalysis.

Figure 2 shows the liquefaction rate of bamboo with different reaction temperature within 90 min and 5 wt.% of catalysis. It was found that fewer residues were obtained at higher temperature and the yield of various compounds products was dependent on the reaction conditions. Therefore, it may be possible to control the structure and properties of liquefied products in the latter stage of liquefaction by controlling reaction time and temperature [1]. However, the condensation reaction had occurred in the high-temperature reaction. Therefore, bamboo liquefaction, application of lower temperature and less amount of additional acid were suggested for preventing the condensation reaction.

3.2 FT-IR analysis

In order to understand the functional groups change in bamboo, liquefied bamboo had been observed by a FT-IR spectrometer. The major functional groups of liquefied products are listed in table 1. Comparing the peak of bamboo with that of its 5 min liquefied residues, a similar peak could be observed. The result indicated that there was still much lignin present in the untreated residues. The FT-IR spectrometer of cellulose as model substance is also shown in fig. 3. The peak was similar with the peak of bamboo 30 min residues. This result indicated that less lignin was contained in residues and proved that acid decomposition lignin was faster than cellulose decomposition. For detailed information, the absorption band from 3500 cm^{-1} to 3100 cm^{-1} was assigned as hydroxide, the band of bamboo LRs at 2908 cm^{-1}, 2907 cm^{-1} was assigned as CH and CH_2 stretching vibration of aliphatic hydrocarbons. Two small peaks at 1743 cm^{-1} and 1736 cm^{-1} were assigned as C=O bending which is attributable to acetyl group or carboxyl group. The band at 1716 cm^{-1} was assigned as C=O which is attributable to carboxyl group stretching conjugated carbonyl group. Two obvious peaks at 1600 cm^{-1} and 1510 cm^{-1} were assigned as C=C bending which is attributable to the aromatic ring of the lignin. Another important band at 1151 cm^{-1} was assigned as C-O-C which is attributable to C-O-C symmetric stretching vibration of lignin β-O-4 ether bond.

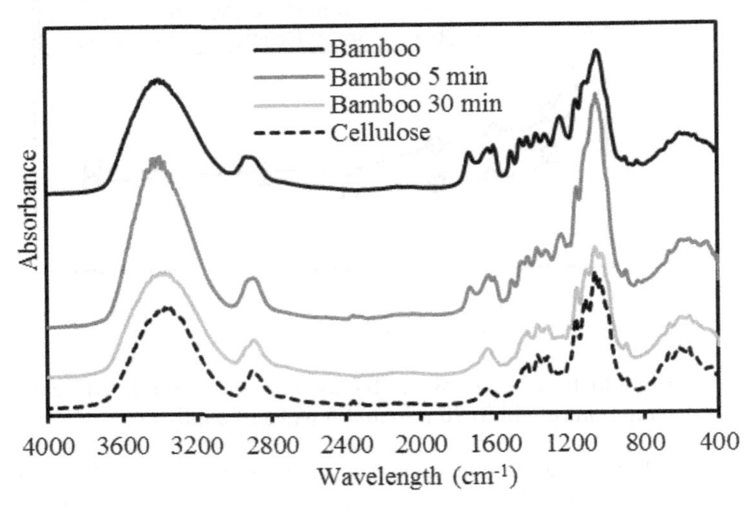

Figure 3: FT-IR spectrometer analysis of liquefied bamboo within 30 min at 120°C.

Table 1: Functional groups of waste bamboo materials and liquefied bamboo.

Absorption band (cm^{-1})	Assignment	Origin from	Reference
3500–3100	O-H	Hydroxide	
2908; 2907	CH; CH_2	CH and CH_2 stretching vibration of aliphatic hydrocarbons	[9]
1743; 1736	C=O	Acetyl group or carboxyl group	
1716	C=O	Carboxyl group stretching conjugated carbonyl group	[10]
1600; 1510	C=C	The aromatic ring of the lignin	[11]
1605; 1508; 1423	C-H	C-H of the aromatic ring	[10]
1430;1370	C-H	It is a group of CH_2	[12]
1151	C-O-C	COC symmetric stretching vibration of lignin β-O-4 ether bond	[9]
1100	C-O	C-O functional group of cellulose	[12]
830	=C-H	Aromatic C-H	

3.3 Surface morphological observation

Fig. 4 shows the SEM image of liquefied bamboo residues (LRs) related to reaction time within 30 min at 120°C.

The samples were raw waste bamboo material, 5 min LRs and 30 min LRs. The different samples, which had been obtained according to reaction time, were

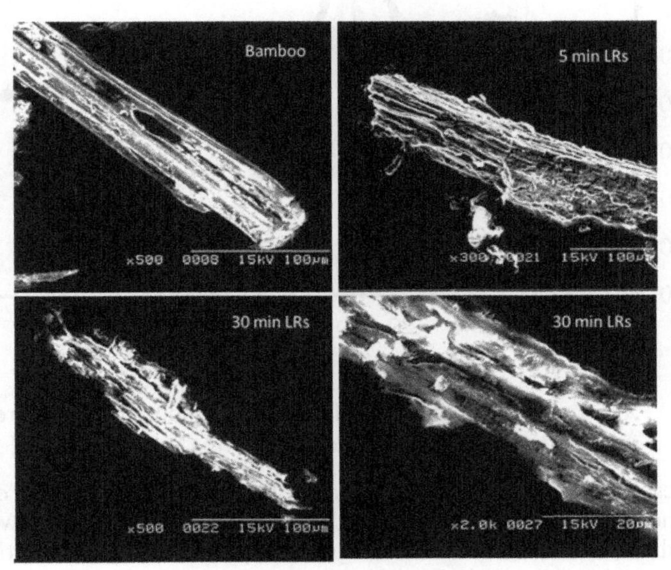

Figure 4: SEM photos of liquefied bamboo residues (LRs) related to reaction time within 30 min at 120°C.

compared during the liquefaction process. Partial destruction was observed in 5 min LRs. The liquefied cellulose particles were not dissolved uniformly as indicated by the SEM image. This randomness made the speed of liquefied bamboo irregular. Most of waste bamboo powder samples had been liquefied. Due to the condensation reaction, the sample was not observed after 30 min.

3.4 X-ray diffraction analysis

X-ray diffraction gives direct results and useful information to explain the structure change of waste bamboo materials and liquefied bamboo. Generally, macro- and micro-structure especially the degree of crystallinity were changed drastically by changing the conditions of liquefaction. In this study, obvious structural changes were found (fig. 5). The apparent crystallinity increased at the beginning of liquefaction. But after 5 min, the apparent crystallinity decreased with the liquefaction time. Such results are not in line with general understanding. Therefore, a model substance was used to explain this phenomenon. Cellulose, lignin and a mixture of cellulose: lignin (2:1) were measured by XRD. It was found that the mixture of cellulose and lignin (2:1) has the lower apparent crystallinity (fig. 6).

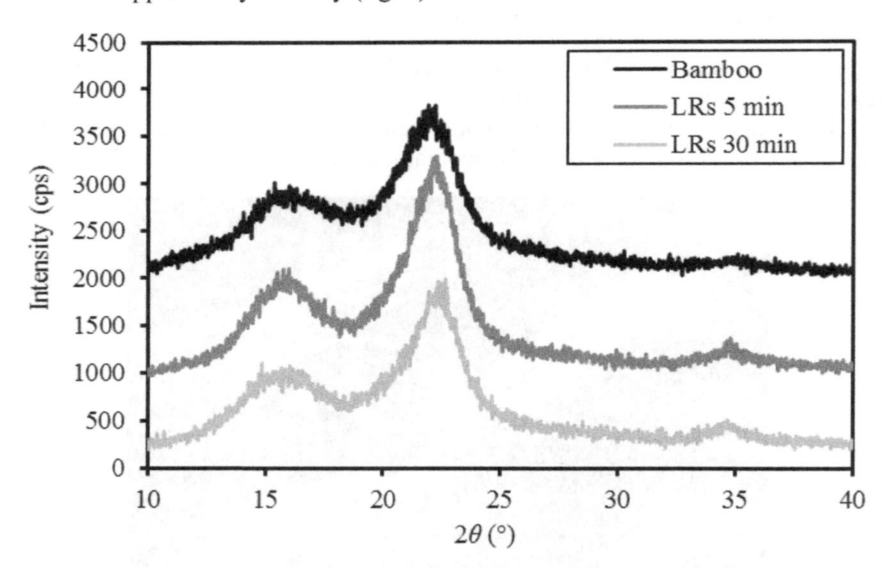

Figure 5: X-ray diffractograms of liquefied bamboo residues related to reaction time within 30 min at 120°C.

This result did not mean that the apparent crystallinity of cellulose decreased because no reaction had taken place during mixture. One reason which can explain the phenomenon is that lignin dispersed x-ray diffraction. The misunderstood result showed the false appearance that the apparent crystallinity had decreased. Therefore, waste bamboo material contains 18.9% lignin before liquefaction. After liquefied, the acid decomposition of lignin was faster than cellulose decomposition [13]. In other words, the peak of LRs 5 min was large because the content of cellulose was increased. Then, the intensity peak of 30 min became weaker. This result indicated that cellulose had been liquefied and the crystalline structure had been broken. Meanwhile, the long chain became short; the degree of polymerization also became low.

The change of apparent crystallinity related to different reaction temperature within 30 min is illustrated in fig. 7. More apparent crystallinity occured at a

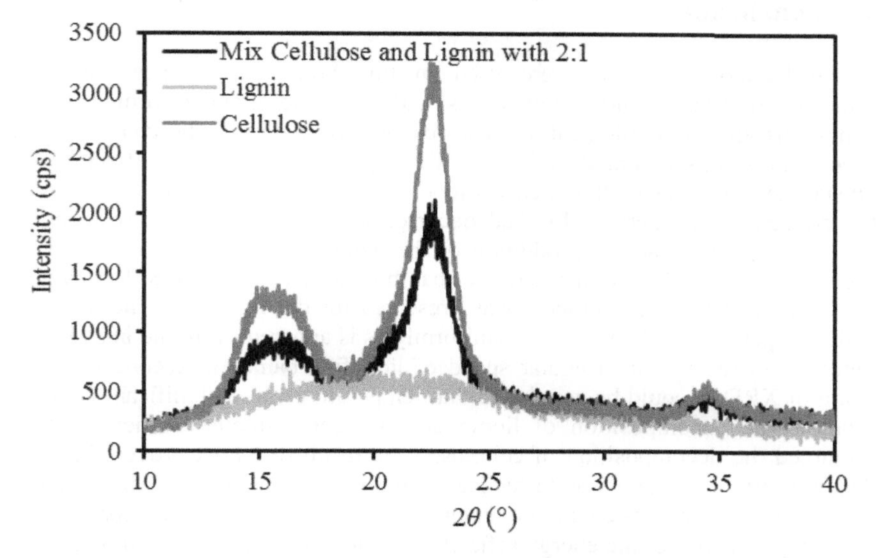

Figure 6: X-ray diffractograms of cellulose and lignin model substance.

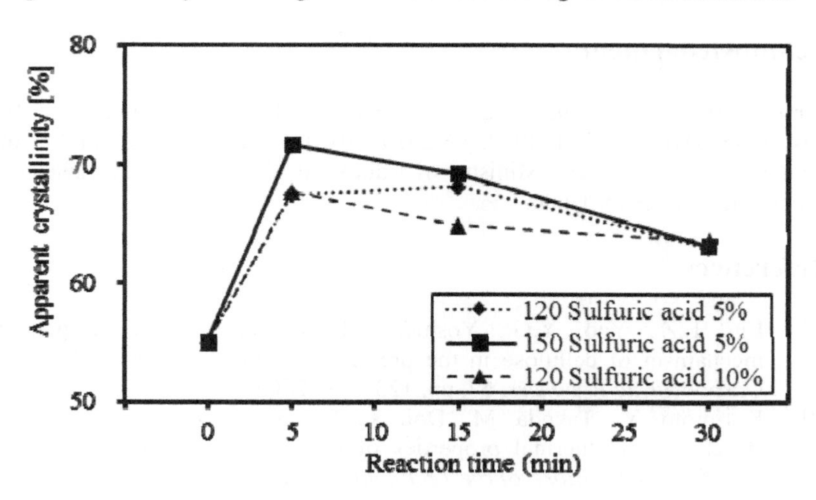

Figure 7: The changes of apparent crystallinity related to different reaction temperature within 30 min.

higher temperature of the liquefaction process. This result suggested that temperature was a significant factor to promote the decomposition of lignin. However, it did not play a significant role in the decomposition of cellulose. If this result is compared with the different amount of acid catalyst, a greater amount of acid catalyst promotes the decomposition of cellulose and also contributes to acid decomposition reaction [14].

4 Conclusion

Waste bamboo materials were used in this liquefaction experiment. The liquefaction process and liquefied residual were measured according to the conditions and morphological surface observed by a SEM. The change in the functional groups was analyzed by a FT-IR spectrometer and the effect on the crystalline structures of liquefied bamboo was determined by XRD. Regarding the condensation reaction, this had occurred in the high-temperature reaction. Lower temperature and less addition of the amount of acid was suggested for preventing the condensation reaction of bamboo liquefaction. The SEM image of the bamboo liquefaction process was presented for the first time. The liquefied cellulose particles did not dissolve uniformly as is apparent from the image. This randomness resulted in irregular speed of liquefied bamboo. According to the result of XRD, it could be concluded that temperature was a significant factor to promote the decomposition of lignin and a greater amount of acid catalyst promoted the decomposition of cellulose. The result of FT-IR proved that acid decomposition of lignin was faster than cellulose decomposition. All the above results suggest that waste bamboo materials can be used for liquefaction but it is necessary to consider the energy efficiency for its high crystalline index leads to a long liquefaction time.

Acknowledgement

Some work of this study was supported by the special funds for Basic Researches (B) (No. 22404022, FY2010~2012) of Grant-in-Aid for Scientific Research of the Japanese Ministry of Education, Culture, Sports, Science and Technology (MEXT), Japan.

References

[1] Lin, L.Z., Yao, Y.G., Yoshida, M. & Shiraishi, N., Liquefaction mechanism of cellulose in the presence of phenol under acid catalysis. *Carbohydrate Polymers*, **57**, pp. 123–129, 2004.

[2] Kurimoto, Y., Takeda, M., Doi, S., Tamura, Y. & Ono, H., Network structures and thermal properties of polyurethane films prepared from liquefied wood. *Bioresource Technology*, **77**, pp. 33–40, 2001.

[3] Lee, W.J. & Chen, Y.C., Novolak PF resins prepared from phenol liquefied Cryptomeria japonica and used in manufacturing moldings. *Bioresource Technology*, **99**, pp. 7247–7254, 2008.
[4] Yip, J., Chen, M., Szeto, Y.S. & Yan, S., Comparative study of liquefaction process and liquefied products from bamboo using different organic solvents. *Bioresource Technology*, **100**, pp. 6674–6678, 2009.
[5] Niu, M., Zhao, G.J. & Alma, M.H., Polycondensation reaction and its mechanism during lignocellulosic liquefaction by an acid catalyst. *Forestry Studies in China*, **13**, pp. 71–79, 2011.
[6] Pan, H., Synthesis of polymers from organic solvent liquefied biomass: A review. *Renewable & Sustainable Energy Reviews*, **15**, pp. 3454– 3463, 2011.
[7] Kobayashi, M., Asano, T., Kajiyama, M. & Tomita, B., Effect of ozone treatment of wood on its liquefaction. *Journal of Wood Science*, **51**, pp. 348–356, 2005.
[8] Zhang, Y.C., Ikeda, A., Hori, N., Takemara, A., Ono, H. & Yamada, T., Characterization of liquefied product from cellulose with phenol in the presence of sulfuric acid. *Bioresource Technology*, **97**, pp. 313–321, 2006.
[9] Lee, W.-J. & Lin, M.-S., Preparation and application of polyurethane adhesives made from polyhydric alcohol liquefied Taiwan acacia and China fir. *Journal of Applied Polymer Science*, **109**, pp. 23–31, 2008.
[10] Izumo, K. & Fukushima, M., Influence of wood species on the properties of biopolyurethane prepared from liquefied wood with residue. *Journal of Applied Polymer Science*, **118**, pp. 2109–2115, 2010.
[11] Chen, F.G. & Lu, Z., Liquefaction of wheat straw and preparation of rigid polyurethane foam from the liquefaction products. *Journal of Applied Polymer Science*, **111**, pp. 508–516, 2009.
[12] Hassan, E.M. & Shukry, N., Polyhydric alcohol liquefaction of some lignocellulosic agricultural residues. *Industrial Crops and Products*, **27**, pp. 33–38, 2008.
[13] Kobayashi, M., Asano, T., Kajiyama, M. & Tomita, B., Analysis on residue formation during wood liquefaction with polyhydric alcohol. *Journal of Wood Science*, **50**, pp. 407–414, 2004.
[14] Sung, P.M. & Jeong, P.J., Liquefaction of cellulose in the presence of phenol using p-toluene sulfonic acid as a catalyst. *Journal of Industrial and Engineering Chemistry*, **15(5)**, pp. 743–747, 2009.

Liquefaction processes and characterization of liquefied products from waste woody materials in different acidic catalysts

Q. Wang[1], Q. Chen[1], P. Apaer[1], N. Kashiwagi[1], H. Kurokawa[1], K. Sugiyama[2], X. Wang[3] & X. Guo[4]
[1]*Graduate School of Science and Engineering, Saitama University, Japan*
[2]*Hachinohe National College of Technology, Japan*
[3]*Center for Environmental Science in Saitama, Japan*
[4]*School of Environmental and Chemical Engineering, Shanghai University, China*

Abstract

The liquefaction process is one of the promising techniques for effective utilization of woody biomass, since the lignocelluloses can be converted to liquid reactive material, as eco-polymeric materials. Japanese cedar (Cryptomeria Japonica), as an abundant waste softwood material, was selected and used in our wood liquefaction experiment. In order to investigate its basic characteristics and potentially harmful metal contents, the composition and metal elements of waste woody samples were determined based on the methods of Japanese Industrial Standard as well as by an ICP-AES, separately. Then the waste woody samples were liquefied by phenol wood liquefaction according to the orthogonal test $L_9(3^4)$, in order to obtain relatively less residue by different reaction conditions. It is thought that sulfuric acid plays an important role in retarding the condensation reaction during the acid-catalyzed phenol liquefaction because of the dehydration, and it can be summarized that the most influential factors of the wood liquefaction conditions were obtained within the setting ranges on four factors and three levels by using the orthogonal tests. In the acidic catalyst comparison experiment, as a result, when using concentrated sulfuric acid as the strong acidic catalyst, the minimum of residual content reached 9.71%. According to these experimental results, the new liquefied samples demonstrated

the relationships between some characteristics of liquefied products from waste woody materials through analyses of Japanese phenolic resin industry testing series such as viscosity, nonvolatility and so on. The results showed that, whether the viscosity or novolatility, the greatest changes of the liquefied materials had taken place during 6 hours in reaction time. In addition, in order to find some structural changes and to clarify the mechanism of respective products, the analysis by combined phenol and free phenol was carried out in further studies.
Keywords: waste woody materials, liquefaction, phenol, liquefied products, orthogonal test, viscosity, acidic catalysts.

1 Introduction

In China and Japan it is estimated that more than a million tones of timber waste is generated each year [1]. Construction woody waste has become the major source of solid organic waste [2]. Waste management is becoming more important against the alarming warning signals in industry and to people's lives. Reuse and recycling to reduce the waste are considered as the only methods to recover this waste, such as timber waste. On the other hand, the polymeric material can be far more seriously damaging in the natural environment, because it is difficult to be degraded in a short period of its life time [3]. The technologies of liquefaction can convert biomass resources into eco-polymeric materials which can replace the pre-polymer which is made from oil resource [4], and solvent liquefaction is one of the effective methods as a solution to the problem [5]. The acid catalyst can degrade the woody components to small ingredients, and then these ingredients react with phenol to form a derivative and dissolve in the liquefaction solvent.

Lin *et al.*, using GG guaiacylglycerol-β-guaiacyl ether, as a lignin model compound, carried out the relevant research on liquefaction reaction mechanism of lignin compounds in the presence of sulfuric acidic catalyst [6, 7] and without catalysts [8, 9]. They also attempted to clarify the reaction mechanism of cellulose with phenol under acid-catalyzed conditions. It was found that the yield ratio of various compounds in the end reaction product is greatly dependent on the reaction conditions, i.e. the phenol/cellobiose ratio, concentration of acidic catalyst, reaction temperature, and reaction time. Therefore, by controlling these reaction conditions, it is possible to adjust the structure and properties of the end liquefaction products [10]. Yamada and Ono [11] reported that the acid catalyzed liquefaction of cellulose with polyhydric alcohols includes complicated reaction where the cellulose is decomposed to glucose, the glucose transforms to 5-hydroxymethylfurfural. However, Ono *et al.* [12] reported a possible chemistry of phenol assisted liquefaction: cellulose is degraded to cello-oligosaccharides and glucose by sulfuric acidic catalyst, then its pyranose ring structure is subject to decomposition and then recombination of its decomposed fragments and phenols would occur in the process of phenol liquefaction. Zhang *et al.* [13] researched the liquefaction mechanism of cellulose in phenol. They indicated that the pyranose, decomposed from cellulose, could combine with phenol to form a hydroxyl benzyl form derivative. Lee and Ohkata [14] indicated wood could be rapidly liquefied at the supercritical temperature. Under this condition, over 90% of wood was liquefied within 0.5 min, as the new liquefaction methods.

In this study, it was found that reaction time is the most important factor in the process of liquefaction according to the result of orthogonal tests, in spite of acidic catalysts used as another important factor. Especially, the extremely rapid and complex changes had taken place during the first 2 hours. A better understanding of the liquefaction process and characterization of liquefied products should be informed through a series of experiments and analysis. In addition, the heavy metal elements such as Cr, Cu and Pb which sourced from chemical preservatives had been tested for the environmental impact assessment, because the liquefaction reaction carried out under strong acidic conditions.

2 Materials and methods

2.1 Materials and different acidic catalysts

The Japanese cedar (Cryptomeria Japonica), as waste woody material from thin woody materials and construction woody materials, was collected from the campus of Saitama University and waste wood processing factories in Kanto region of Japan. The air-dried samples milled in smashing equipment and retained at size ranges of 10–100 mesh screens were used for the test. All the flour samples of woody materials were dehydrated in an oven at 107°C for 24 hours. The main different acidic catalysts such as H_2SO_4 (95%), H_3PO_4 (85%), phenol and other reagents were prepared with analytical grade reagents in accordance with the Japanese industrial standard (JIS) supplied by Wako Pure Chemicals, Co. Ltd., Japan.

2.2 Chemical composition of liquefied woody materials

The composition analysis of liquefied woody material was divided into three parts. In the industry analysis, the test of ash, which is the residue powder left after the combustion of wood, is performed according to JIS-M8812. The moisture content of wood is determined by measuring the weight loss after drying the sample at 105°C in an oven. Moisture content is determined on a separate portion of the sample not used for the other analyses. Volatile matter is tested by heating the wood sample to 900°C under carefully controlled conditions and measuring the weight loss. The fixed-carbon contents in woody materials were determined by subtracting the percentages of moisture, volatile matter and ash from each sample. The second part of analysis was elemental analysis, which was carried out according to JIS-M8813 by the CHN coder (Model MT-5, Yanaco Co. Ltd., Japan). In the part of component quantitative analysis, wood extraction procedures in 1% NaOH extract most extraneous components, some lignin and low molecular weight hemicelluloses and degraded cellulose. The solubility of woody material in EtOH/benzene (benzene is a known carcinogen; toluene can be substituted) in a 1:2 volume ratio will give a measure of the extractives content. The wood meal is refluxed 6 hours in a soxhlet flask, and the weight loss of the extracted, dried woody material is measured. Holocellulose is the total polysaccharide (cellulose and hemicelluloses) contents of woody material and methods for its determination seek to remove all of the lignin from woody material without disturbing the carbohydrates. Alpha cellulose is obtained after treatment of the holocellulose with 17.5% NaOH. Cross and bevan cellulose consists largely of pure cellulose, but also contains some hemicelluloses. It was obtained by chlorination of wood meal, followed by washing with 3% SO_2 and 2% Sodium Sulfite. The lignin contents of waste woody materials presented in the tables are Klason lignin, the residue remaining after solubilizing the carbohydrate with strong mineral acid. The usual procedure was to adopt a shorter method, which treats the sample with 72% H_2SO_4 at 30°C for one hour, followed by an hour at 120°C in 3% H_2SO_4. The analytical results of the waste woody materials used in our experiments were summarized in table 1 and table 2.

Table 1: Industrial and elemental analysis (wt. %) of waste woody materials.

Analysis of waste woody materials				Elements analysis			
Ash	VM	M	FC	C	H	N	O
0.6	79.1	7.7	12.6	48.9	6.1	0.4	44.6

VM: Volatile Matter, M: Moisture, FC: Fixed Carbon.

Table 2: Composition analysis (wt. %) of waste woody materials.

Composition analysis		Liquefied woody materials
Carbohydrate	Holocellulose[a]	55
	Cross and Bevan Cellulose[b]	43
	Alpha Cellulose[c]	29
	Klason Lignin	27
Solubility	1%NaOH	12.81
	EtOH/Benzene	3.25

[a]Holocellulose is the total carbohydrate content of waste woody materials.
[b]Cross and Bevan Cellulose is largely pure cellulose but contains some hemicelluloses.
[c]Alpha Cellulose is nearly pure cellulose.

2.3 Heavy metal analysis in thin and construction woody materials by an ICP-AES

The heavy metals such as Cr, Cu, and Pb which are sourced from wood material were tested, because the liquefaction reaction is carried out in strong acidic conditions. The inductively coupled plasma atomic emission spectroscopy (ICP-AES Optima 5300 DV, Perkin Elmer Co., Ltd., Japan) analysis was used as the analysis tool, the two kinds of woody material were digested with concentrated HNO_3 (3 ml) and HF (3 ml), then the solutions were heated to dryness on an aluminum heating block at 100°C. After 1 h adding HNO_3 (10 ml) and HF

(2 ml), these extract solutions were left overnight on a heating block, and the temperature raised slowly to 200°C and heated to dryness. The residue in the tubes was then leached with 0.1 N of HNO_3 and made-up to final volumes of 50 ml sample to analyse the heavy metal ions.

2.4 The orthogonal test L_9 (3^4) using in the liquefaction of woody materials

An orthogonal test L_9 (3^4) design was used for optimizing the extraction conditions of liquefaction of waste woody materials in the presence of sulfuric acidic catalysts. In this study, the setting conditions were carried out at reaction temperatures of 140°C, 150°C and 160°C, the ratio of phenol and air-dried waste woody materials were 1:3.0, 1:3.5 and 1:4.0, extraction time 1.5, 2, 2.5 and additive weight of catalysts 6, 8 and 10 (wt% of solvent), respectively, on the basis of the single-factor test. Table 3 shows the experimental conditions for the extraction of liquefaction. It is according to the other experimental conditions, which are usually used in the experimental wood liquefaction. It was thought that high temperature will affect the volatile of phenol and the lower temperature will affect the speed of experimental liquefaction. Another experimental factor will follow an interrelated factor such as the selected liquefaction temperature. In this orthogonal test, it is hoped that the better among these factors and the optimal among setting conditions would be identified.

Table 3: Experiment factors and levels for orthogonal test.

Variable	Level 1	Level 2	Level 3
A: Extraction temperature (°C)	140	150	160
B: Extraction time (hour)	1.5	2.0	2.5
C: Ratio of phenol and air-dry wood	1:3.0	1:3.5	1:4.0
D: Additive weight of catalysts (wt %)	6	8	10

2.5 Characteristics of phenol liquefaction of woody materials

2.5.1 Experimental comparison of different acidic catalysts

Experimental liquefaction of waste woody materials were carried out by varying the weight charge ratio of wood to phenol with 1/4 and the reaction temperature at 150°C by oil bath into a 500 ml three necked flask equipped with stirrer. In addition, the same weight ratio of wood/phenol/H_2SO_4, HCl and H_3PO_4 were used in the comparison experiments. The content was heated for 2 hours.

Otherwise, the experimental research product was liquefied for 6 hours with two kinds of acidic catalysts; the residues, viscosity, nonvolatility, combined phenol and free phenol were tested at every hour to evaluate the correlation.

2.5.2 Measurement of liquefaction residue

The liquefied woody material was weighted and diluted with methanol, and Whatman filter paper was used to collect the liquefied residue. The liquefied residue was dried in a heating oven at 105°C in 24 hours and the residues were calculated by equation

$$\text{Liquefied residue (wt \%)} = (\text{Residue /Raw material}) \times 100\% \quad (1)$$

2.5.3 Viscosity measurement of liquefied waste woody materials

The viscosity of liquefied woody materials was measured with a Viscometer (Model VT-04F, RION Co. Ltd., Japan) by each hour within 6 hours in 300 ml standard beaker at 25±3°C. The viscosity is measured using a special mechanism to obtain direct readings in decipascal-seconds (dPa.s) (accuracy: within ±10% of meter indication, rotating velocity of the rotor: 62.5 rpm).

2.5.4 Nonvolatility in the liquefied waste woody materials

About (1.5-3.0)±0.5 g of liquefied waste woody materials were weighted and dried using a oven at 180±1°C by 1 hour according to JIS-K6910 to remove the un-reacted phenol and nonvolatility in the liquefied waste woody materials can be calculated by the following equation

$$\text{Nonvolatility (wt \%)} = (\text{Residue after oven-dry /The sample}) \times 100\% \quad (2)$$

2.5.5 Combined phenol and free phenol in the liquefied woody materials

The combined phenol and free phenol were used to show the phenol existing in the liquefied waste woody materials that were combined with wood components or free in the system [5], respectively. Both the combined and free phenol was calculated from the nonvolatile content in the liquefied wood. The combined phenol was calculated by subtracting the content of wood used in the liquefaction system from the nonvolatile content after liquefaction. The free phenol (unbound phenol) was calculated by subtracting the content of combined phenol from the content of phenol used in the initial liquefaction system. The combined phenol can be obtained from

$$\text{Combined phenol (wt \%)} = [\text{Nonvolatility} - w/(w + p + c)] \times 100\% \quad (3)$$

and the free phenol from

$$\text{Free phenol (wt \%)} = [p/(w + p + c) - \text{Combined phenol}] \times 100\% \quad (4)$$

where w is the weight of woody material, p the phenol weight and c the catalyst weight.

3 Results and discussions

3.1 Comparison of heavy metal analysis between thin woody material and construction woody material (Liquefied woody materials)

Analytical results of the heavy metal concentrations in the thin woody materials and construction woody materials by ICP-AES analysis are shown in fig. 1. It shows that the arithmetic means of Co, Cr, Cu, Ni and Pb are 0.02, 0.04, 0.14, 0.04 and 0.09 mg g^{-1} of sample, respectively, both of woody material have the nearly concentrations of these heavy metals. These metals are a cause of environmental pollution; meanwhile, an amount of concentrated acidic catalyst was used in the liquefaction process; the reaction between the catalyst and heavy metals, which may come from wood preservatives, was considered in the liquefaction system.

Klok and de Roos [15] showed that heavy metal contamination can induce two major effects on the ecosystem level: (a) accumulation of e.g. Cr can lead to risks of secondary poisoning, while (b) organism disappears even at low levels of Cu in soil. The conditions would occur if the product of liquefied woody materials would be landfilled in soil. In order to further understand the state of these metals after liquefaction, ion-exchange test will be carrying out to find whether harmful ions (Pb^{2+}, Cd^{2+} and Ni^{2+}) would be produced.

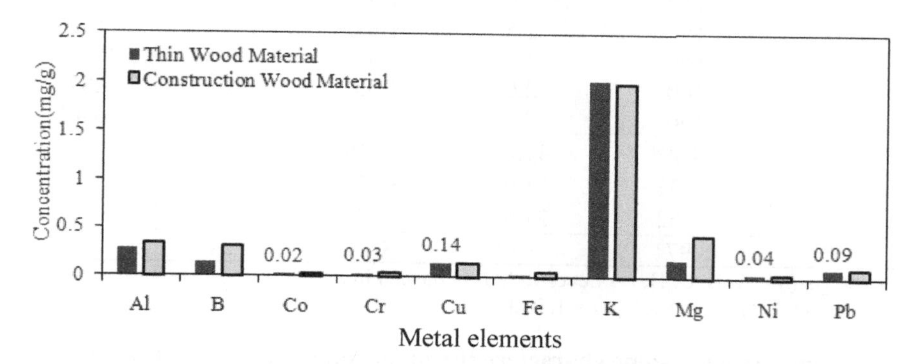

Figure 1: Some metal elements in thin and construction woody materials.

3.2 Influential factors of liquefaction examined by the orthogonal tests

In this study, all influential factors were examined using an orthogonal L_9 (3^4) test design. The total evaluation indexes were used in the analysis by the statistical method. The results of the orthogonal test and extreme difference analysis are presented in table 4.

The liquefied residue in the liquefied products from each test was weighted. As shown in table 4, the influence of the compounds on the mean liquefied residue decreases in the order: B (extraction time) > A (extraction temperature) > C (ratio of phenol and wood) > D (catalyst weight) according to a comparison of the *R* values.

The extraction time was found to be the most important factor of the yield. Moreover, the minimum residue of the liquefied products can be obtained when extraction time, extraction temperature, phenol to wood ratio and catalyst weight are 2 hour, 150°C, 1:4 and 8%, respectively. However, the best extraction conditions cannot be obtained based only on the outcomes of table 4 although, in summary, extraction time is the most influential factor under wood liquefaction conditions among the setting ranges of four factors and three levels using the orthogonal tests. The tangible results provided us with the correct catalyst using the orthogonal tests, and this should be in comparisons of some characteristics of the products in further experiments.

Table 4: The scheme and analytical values of L_9 (3^4) orthogonal test.

Sample No.	A: extraction temperature (°C)	B:extraction time (hr)	C: ratio of phenol and wood	D: catalyst weight (wt %)	Liquefied residue (wt %)
1	140	1.5	1:3.0	6	17.49
2	140	2.0	1:3.5	8	10.83
3	140	2.5	1:4.0	10	10.27
4	150	1.5	1:3.5	10	14.28
5	150	2.0	1:4.0	6	11.20
6	150	2.5	1:3.0	8	11.83
7	160	1.5	1:4.0	8	9.27
8	160	2.0	1:3.0	10	9.87
9	160	2.5	1:3.5	6	10.48
K_1	38.59[a]	41.04	39.19	39.17	
K_2	37.31	31.90	35.59	31.93	
K_3	29.62	32.58	30.74	34.42	
R	8.97[b]	9.14	8.45	7.24	

[a] $K_i^n = \sum n_i$ (n = A, B, C, D; i = Level 1, 2, 3) in dashed line
[b] $R = \max\{ K_i^n \} - \min\{ K_i^n \}$ (i = 1, 2, 3)

3.3 Comparison of some characteristics of phenol liquefaction of wood

Three kinds of catalyst were chosen in the liquefaction experiments to determine the difference they make to liquefaction process. Fig. 2 shows that through the use of a concentration of 95% sulfuric acid, the reaction uses 10% of the total weight of the reactants and the residual substance is at least 9.71%.

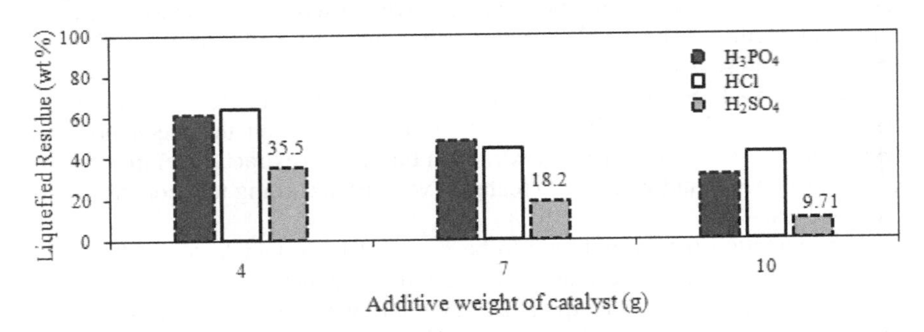

Figure 2: Results of different concentrations of the different acidic catalysts in the liquefaction experiments.

With hydrochloric acid as liquefaction catalyst, the residue rate increased as the amount of catalyst is decreasing. However, due to hydrochloric acid being highly volatile and the liquefaction test carried out under atmospheric pressure, hydrochloric acid was at high temperature under the action of hydrogen chloride gas in the form of evaporation [16]. The product residue rate of 40% is relatively higher than that for phosphoric acid and sulfuric acid. When using phosphoric acid as catalyst, the results showed that under the same temperature conditions, the use of phosphoric acidic catalyst in the liquefaction of wood has the significant effect of residue increase as the amount of catalyst decreases. However, when the catalyst dosage is greater than 7%, then the extent of catalytic effect declines.

3.4 Characteristics of phenol liquefaction of woody materials with two different acidic catalysts

In order to understand the characteristics of produce for different catalysts, the liquefaction experiment was performed for 6 hours with two kinds of catalysts; the residue, viscosity, nonvolatility, combined phenol and free phenol were tested at every hour to evaluate the correlation. The results are shown in table 5, fig. 3 and fig. 4. Fig. 3 shows the properties of phenol liquefied Japanese cedar. The liquefied wood that used H_2SO_4 as an acidic catalyst had the unliquefied wood residue lower than that when H_3PO_4 was used as an acidic catalyst. This result indicated that Japanese cedar liquefied with H_2SO_4 as a catalyst had a better liquefaction reaction than with H_3PO_4. Nevertheless, the former had a viscosity (950-1000 dPa.s) higher than that of the latter (90-100 dPa.s) in table 5.

Table 5: Viscosity of liquefied products that used H_2SO_4 and H_3PO_4 catalysts over 6 hours.

Time (hour)	1	2	3	4	5	6
Viscosity of produce in H_2SO_4 [a] (dPa.s)	20–23	135–140	250–260	440–460	600–650	950–1000
Viscosity of produce in H_3PO_4 [b] (dPa.s)	8–9	23–25	35–37	60–65	85–90	90–100

[a]The weight ratio of phenol/wood/ H_2SO_4 used for liquefaction was 4/1/0.08.
[b]The weight ratio of phenol/wood/ H_3PO_4 used for liquefaction was 4/1/0.19.

The value would be nearly 10 times higher which showed at a reaction time of 6 hours. A large amount of H_3PO_4 catalyst had been used in the experimental liquefaction shown in fig. 3.

From these results, it can also be found that the greatest changes in the viscosity of liquefied products had taken place in reaction time during the 6 hours of using H_2SO_4. However, small changes in the viscosity of liquefied products occur using H_3PO_4 as the acidic catalyst. During the first 3 hours, the effect of the acidic catalyst is not very clear in the case of H_3PO_4 which could be understood from the results given in fig. 3.

After liquefaction, the nonvolatile contents in the liquefied products were 85.11% for that using H_2SO_4 in the six hour and 57.34% for that using H_3PO_4. These higher values of nonvolatility might be caused by some of the phenol that would combine with the wood component to form some derivatives [5].

In summary, these experiments showed that the nonvolatility value increased with time for H_3PO_4, increased with time when H_2SO_4 was used as a catalyst.

However, the values of combined phenol and free phenol were not calculated at the first hour for H_3PO_4, because many residues also remain after liquefaction which may be unliquefied woody material in the original samples. Zhang *et al.* [13] indicated that the cellulose would be degraded to oligosaccharide and glucose in phenol with an acidic catalyst under high temperature. The pyranose ring would decompose further and combine with phenol to form their derivatives. The derivatives formed had the hydroxyl benzyl group but not the phenoxyl–form because the phenolic functional group that kept in the derivatives had a similar reactivity as phenol. In our study, the combined phenol with H_2SO_4 is better than using H_3PO_4, because more free phenol will be residues in the liquefied products after liquefaction.

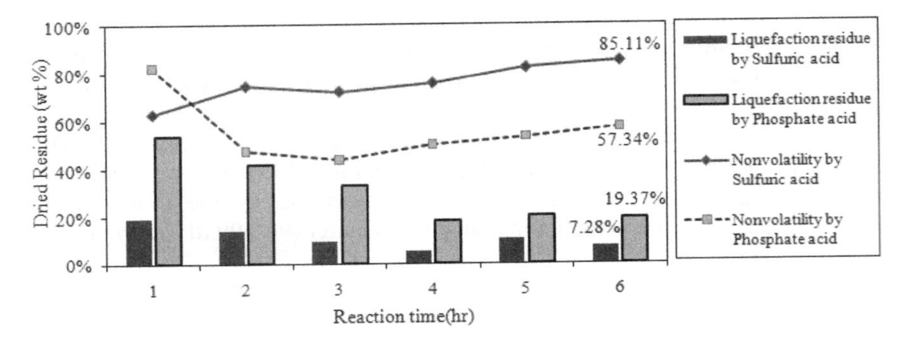

Figure 3: Relationship between liquefaction residues and nonvolatility in liquefied products that used H_2SO_4 and H_3PO_4 in 6 hours (wt. %).

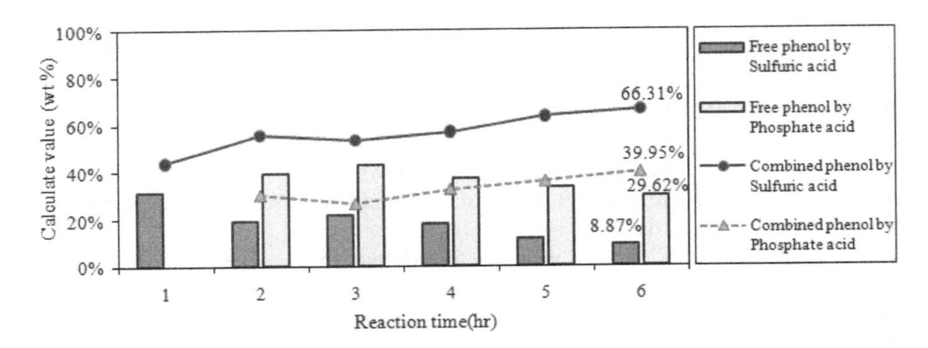

Figure 4: Calculate values of combined phenol and free phenol that used H_2SO_4 and H_3PO_4 in 6 hours (wt. %).

4 Conclusion

The heavy metal concentrations in the thin woody materials and construction woody materials were investigated. The analytical results showed that some heavy metals are contained in wood such as Co, Cr, Cu, Ni and Pb. The possible sources of these metals may be preservatives, paint or other chemicals which were used in wood processing. Since now a common way of waste disposal is landfill, it is necessary to consider that the concentrations of heavy metals would take the reaction with the various catalysts, and the concentrations should be determined before liquefaction. Otherwise, as new material products, the standard amounts of metals contained should be established before use in order to avoid the soil contamination when it would be landfill in the soil.

The results of orthogonal test show that the maximum impact factors can be obtained in the setting range value on four factors and three levels. Thus, it is concluded that it is possible to find the maximum or minimum environmental impact factors in liquefaction process by the orthogonal test. However, in this study, the best extraction conditions by the orthogonal test cannot be obtained, because some other impact factors such as stirring speed will also affect the results. Therefore, seeking the best conditions by only the orthogonal test cannot be recommended. On other words, the minimum residual value in liquefied products is sought, and the orthogonal test can help to ask for and choose better factors in fewer experiments.

The effect of sulfuric acid as the acidic catalyst is obviously better than that of phosphoric acid, which could be known by some properties of viscosity, nonvolatility, combined phenol and free phenol. This may be due to the role of sulfuric acid as a strong acid, but the performance of the liquefied products needs to be analyzed in further studies. In other words, it cannot be said that this kind of acidic catalyst is most suitable for the liquefaction of woody materials; the performance of their products will also decide the catalyst.

Acknowledgements

Some work in this study was supported by the Fund of the FY2009 Research Project of Innovative Research Organization, Saitama University, Japan. The authors also acknowledge the Japanese Ministry of Education, Culture, Sports, Science and Technology (MEXT) of Japan for providing the Special Funds for Basic Researches (B) (No. 19404021, FY2007~FY2009 and No. 22404022, FY2010~2012) of Grant-in-Aid for Scientific Research.

References

[1] Lu, R.S. & Hua J., A vital way of saving wood: utilization of china urban waste wood. *China Wood Industry*, **20**, 2006 (in Chinese).

[2] Wang, Q., Endo, H., Shukuzaki, N., Sekiguchi, K., Sakamoto, K., Kurokawa, H., Nakaya, Y. & Akibayashi, T., Study on char–biomass briquette of pyrolyzed materials from industrial organic wastes. *Proc. of Int. Conf. of Renewable Energy 2006*, P–B–25, pp. 1135–1140, 2006.

[3] Yoshida, C., Okabe, K., Yao, T., Shiraishi, N. & Oya, A., Preparation of carbon fibers from biomass–based phenol formaldehyde resin. *Journal of Materials Science*, **40**, pp. 335–339, 2005.

[4] Behrendt, F., Neubauer, Y., Oevermann, M., Wilmes, B. & Zobel, N., Direct liquefaction of biomass. *Chemical Engineering Technology*, **31(5)**, pp. 667–677, 2008.
[5] Lee, W.J. & Chen, Y.C., Novolak PF resins prepared from phenol liquefied Cryptomeria japonica and used in manufacturing moldings. *Bioresource Technology*, **99**, pp. 7247–7254, 2008.
[6] Lin, L.Z., Yao, Y. & Shiraishi, N., Liquefaction mechanism of β–O–4 lignin model compound in the presence of phenol under acid catalysis. Part 1. Identification of the reaction products. *Holzforschung*, **55**, pp. 617–624, 2001a.
[7] Lin, L.Z., Yao, Y. & Shiraishi, N., Liquefaction mechanism of β–O–4 lignin model compound in the presence of phenol under acid catalysis. Part 2. Reaction behaviour and pathway. *Holzforschung*, **55**, pp. 625–630, 2001.
[8] Lin, L.Z., Yoshioka, M., Yao, Y. & Shiraishi, N., Liquefaction mechanism of lignin in the presence of phenol at elevated temperature without catalysts. I. Structural characterization of the reaction products. *Holzforschung*, **51**, pp. 316–324, 1997.
[9] Lin, L.Z., Yoshioka, M., Yao, Y. & Shiraishi, N., Liquefaction mechanism of lignin in the presence of phenol at elevated temperature without catalysts. II. Reaction pathway. *Holzforschung*, **51**, pp. 325–332, 1997.
[10] Lin, L.Z., Yao, Y., Yoshioka, M. & Shiraishi, N., Liquefaction mechanism of cellulose in the presence of phenol under acid catalysis. *Carbohydrate Polymers*, **57**, pp. 123–129, 2004.
[11] Yamada, T. & Ono, H., Characterization of the products resulting from ethylene glycol liquefaction of cellulose. *Journal of Wood Science*, **47(6)**, pp. 458–464, 2001.
[12] Ono, H., Zhang, Y.C. & Yamada, T., Dissolving behavior and fate of cellulose in phenol liquefaction. *Transactions of the Materials Research Society of Japan*, **26(3)**, pp. 807–812, 2001.
[13] Zhang, Y.C., Ikeda, A., Hori, N., Takemura, A., Ono, H. & Yamada, T., Characterization of liquefied product from cellulose with phenol in the presence of sulfuric acid. *Bioresource Technology*, **97(2)**, pp. 313–321, 2006.
[14] Lee, S.H. & Ohkata, T., Rapid wood liquefaction by supercritical phenol. *Wood Science and Technology*, **37**, pp. 29–38, 2003.
[15] Klok, C. & de Roos, A.M., Population level consequences of toxicological influences on individual growth and reproduction in Lumbricus rubellus (Lumbricidae, Oligochaeta). *Ecotoxicology and Environmental Safety*, **33**, pp. 118–127, 1996.
[16] Alma, M.H., Yoshioka, M., Shiraishi, N. & Yao, Y., Some characterizations of hydrochloride acid catalyzed phenolated wood–based materials. *Mokusai Gakkaishi*, **41(8)**, pp. 741–748, 1995.

The anaerobic digestion of cattle manure: the effect of phase-separation

V. Yılmaz[1] & G. N. Demirer[2]
[1]*Akdeniz University, Turkey*
[2]*Middle East Technical University, Turkey*

Abstract

Various aspects of anaerobic digestion (AD) technology have been the focus of research in recent years. Shortening the digestion time with enhanced process efficiency is one of the integral concerns in AD technology. This study was conducted to investigate the feasibility of a two-phase anaerobic treatment system for unscreened dairy manure. Hydraulic retention time (HRT) and organic loading rate (OLR) in the hydrolytic reactor are varied in order to evaluate the effect of these factors. The results showed that an optimum HRT and OLR of 2 days and 15 g.VS/L.day, respectively, yielded maximum acidification. The separation of acidogenic and methanogenic phases of digestion resulted in a significant increase in methane production rate in the methane reactor. The methane yields were found to be 313 and 221 mL CH_4/g.VS loaded in two-phase and one-phase systems at 35°C, respectively.
Keywords: anaerobic digestion, dairy manure, two-phase, methane.

1 Introduction

With the rapid depletion of conventional energy sources, the need to find alternative, but preferably renewable, sources of energy is becoming increasingly acute. Through anaerobic digestion of biomass, including animal wastes, useful energy can be obtained [1]. Biogas plants are expected to be an effective solution to the manure management problem providing benefits such as energy saving, environmental protection and reduced CO_2 emissions.

Anaerobic digestion of organic matter became more and more attractive in the recent past because new reactor designs significantly improved the reactor performance [2]. Studies have shown that anaerobic treatment is a stable process under proper operation. But parameters such as process configuration, temperature, biomass, pH, nutrient, and substrate must be carefully scrutinised in order to make successful anaerobic treatment. Many process configurations have been investigated. An improvement in the efficiency of anaerobic digestion can be brought about by either digester design modification or advanced operating techniques [3].

On farms anaerobic digestion of animal manure is an attractive technique for both energy and organic fertilizer production. The literature on manure digestion is mainly focused on liquid manure (i.e. total solids<100 g/l) digestion. Nevertheless many farms, especially smaller ones, throughout the world, still produce solid manure. For on-farm application the digestion system should be as simple as possible to operate and in agreement with the on-farm practice [4].

Conventional anaerobic digestion is proceeded in a single reactor where acidogenesis and methanogenesis both occur. Acidogenesis and methanogenesis are respectively proceeded in two separate reactors and each phase is in the best environmental conditions [3]. This phase separation can be achieved by maintaining a very short HRT in the acid phase reactor. The effluent from the first, acid-forming, phase is then used as the substrate for the methane-phase reactor [5].

One relevant feature of the two-phase approach is that when a high solid containing waste is introduced to the first phase it is liquefied along with acidification. This translates into less liquid addition and, thus, less energy requirements for heating, storing and spreading for two-phase AD systems. The results of several studies [6–14] have clearly demonstrated the applicability and efficiency of two-phase AD for high solids containing wastes.

The advantages of two-phase operation have been extensively documented [15, 16]. Prospects for the phased anaerobic treatment of wastewater are promising. With the variety of reactor designs available and the amenability of reactors to modification, existing treatment systems may be replaced or upgraded as required to achieve increased stability, higher loading capacities and greater process efficiencies than are possible using single-stage systems [3]. Even though several aspects of two phase configuration including liquefaction might be very significant for efficient AD of dairy manure, its application has been limited to screened dairy manure only [1, 17, 18].

This work aimed to evaluate the feasibility of a two-phase anaerobic treatment system for unscreened dairy manure. The specific objective was to compare the effects of different HRT and OLR for optimum acidification.

2 Materials and methods

Wet manure was collected from a private dairy around Gölbaşı, Ankara, and stored at 4°C prior to use. The composition of the dairy manure used in this study had the following characteristics; total solids (TS), 20.1 ±1.7%, volatile solids (VS), 67 ± 4.6% of TS and density, 1042 ± 0.04 g/L. The raw manure was diluted with water to decrease the solids content to achieve slurry with 3.5 and

15 g.VS/L. The relationship between chemical oxygen demand (COD) and VS of this manure was found as 1.04.

The mixed anaerobic culture used as seed was obtained from the anaerobic sludge digesters at the Ankara wastewater treatment plant, which has a solids retention time (SRT) of 14 days. The mixed anaerobic culture was concentrated by settling before being used as inoculum. The volatile suspended solids (VSS) concentration of the concentrated seed cultures used was 23930 ± 3162 mg/L.

2.1 Experimental set-up

In the first part of the study, the optimum retention time and organic loading rate (OLR) values leading to maximum acidification and VS reduction were investigated. Thus, nine daily-fed continuously-mixed acidogenic anaerobic reactors with no recycle were operated as duplicates. The experiments were performed in 250 mL serum bottles capped with rubber stoppers. The reactor operation involved daily feeding of wet dairy manure and wasting of the corresponding reactor contents as indicated in table 1. Solids and hydraulic retention times (SRT/HRT) applied to each reactor were the same since no recycle of the effluent was practiced. Initially, each reactor was seeded with 100 mL of concentrated anaerobic seed cultures. The next day dairy manure (25 mL to reactors 1–3, 50 mL to reactors 4–6, and 80 mL to reactors 7–9) were added to each reactor. Daily feeding and wasting were conducted as seen in table 1. The reactors were flushed with N_2/CO_2 gas mixture for 3 min and maintained in an incubator shaker at 35 ± 1°C and 165 rpm.

Table 1: Daily feeding and wasting used for acidogenic reactors.

Reactor	SRT (days)	OLR (g.VS/L day)	Volume of feeding/wasting (ml)
1	4	5	25
2	4	10	25
3	4	15	25
4	2	5	50
5	2	10	50
6	2	15	50
7	1.25	5	80
8	1.25	10	80
9	1.25	15	80

The one-phase conventional configuration (R1) was run as the control for the two-phase configuration (R2). The effective volumes of R1, R21, and R22 were 1000, 400, and 1000 mL, respectively. The two-phase configuration contained R21 and R22 as the first (acidogenic) and second (methanogenic) phases.

The SRT/HRT values of R1, R21, R22 and the overall two-phase configuration were 20, 2, 8.6, and 10.6 days, respectively. The gas production in R1, R21 and R22 was monitored by a water replacement device. One set of reactors were maintained at 25°C in a temperature-controlled water bath and the others at 35°C (±2) in a controlled room, and all reactors were shaken manually once daily after conducting the gas production measurement. R1, R21, and R22 were seeded with 500, 200, and 500 mL of mixed anaerobic seed culture. The performance of the reactors was monitored by measuring biogas production and soluble COD, VS, volatile fatty acid (VFA), and pH.

2.2 Analytical methods

The pH, daily gas production, total solids, volatile solids, methane percentage, total volatile fatty acids (TVFA) and effluent soluble COD (sCOD) were monitored in each reactor. pH, TS, VS analysis was performed using Standard Methods [19]. sCOD was measured using Hach COD vials according to the EPA approved digestion method [20]. Accordingly, after 2 h digestion, sCOD of sample were directly read using Hach 45 600-02 spectrophotometer (Hach Co. Loveland, Co., USA). TVFA and biogas composition were measured by gas chromatography as described by Yilmaz and Demirer [21].

3 Results and discussion

Nine acidogenic anaerobic reactors were operated for 57 days to determine the optimum SRT and OLR values resulting in maximum acidification and in turn VS reduction. Three different OLRs (5, 10 and 15 g.VS/L.day) were applied to the reactors. For each OLR value, three SRTs (1.25, 2 and 4 days) were studied (table 1). The results are given in fig. 1 in terms of the change in the operating parameters (pH, TVFA, VS, cumulative gas production (CGP), methane content and sCOD) with respect to the combination of OLR and SRT values. Figure 1 does not include the data points within the first "3×SRT" days (12 days for R1–R3, 6 days for R4–R6, and 4 days for R7–R9) which are the theoretical time to reach steady-state conditions in a continuous reactor.

As seen in fig. 1a, pH drop was inversely proportional to the increase in the SRT for each OLR studied. Similarly, for each SRT studied, as the OLR increased, pH decreased. It was observed that the extent of pH drop increased with the increase in the OLR being smallest for the lowest OLR of 5 g.VS/L.day. Besides, it should be noted that the extent of pH drop was also affected by the SRT. For all the OLRs studied, the extent of pH drop for the SRT increase from 1.25 to 2 days was greater than that observed for SRT of 2 to 4 days. It is a well known fact that low retention times and high loading rates lead to higher acidification in two-phase systems. However, as seen in fig. 1a, average pH values observed in the reactors were within 6.2–6.6 and the extent of pH drops was lower relative to acidification of other high solid substrates such as organic fraction of municipal solid wastes. Han *et al.* [22] operated the MUSTAC (multi-step sequential batch two-phase anaerobic composting) process to recover methane and composted material from food waste, where the pH ranged between 6.5 and 7.0 during acidogenic fermentation step. In another research, Kübler and Schertler [23] demonstrated that the favourable pH condition was 6.7 in the three-phase anaerobic degradation of solid waste. Verrier *et al.* [24] stated that both mesophilic and thermophilic liquefaction and acidogenesis of vegetable solid wastes were found to be maximal when the pH was maintained at approximately 6.5 in the hydrolysis reactor. The relatively high pH values observed in this study can be explained by the alkalinity generated by the anaerobic biodegradation of nitrogenous organic compounds contained in the dairy manure used [25, 26]. Similar self-buffering capacity of the manure was also observed in other acidification studies [12, 24].

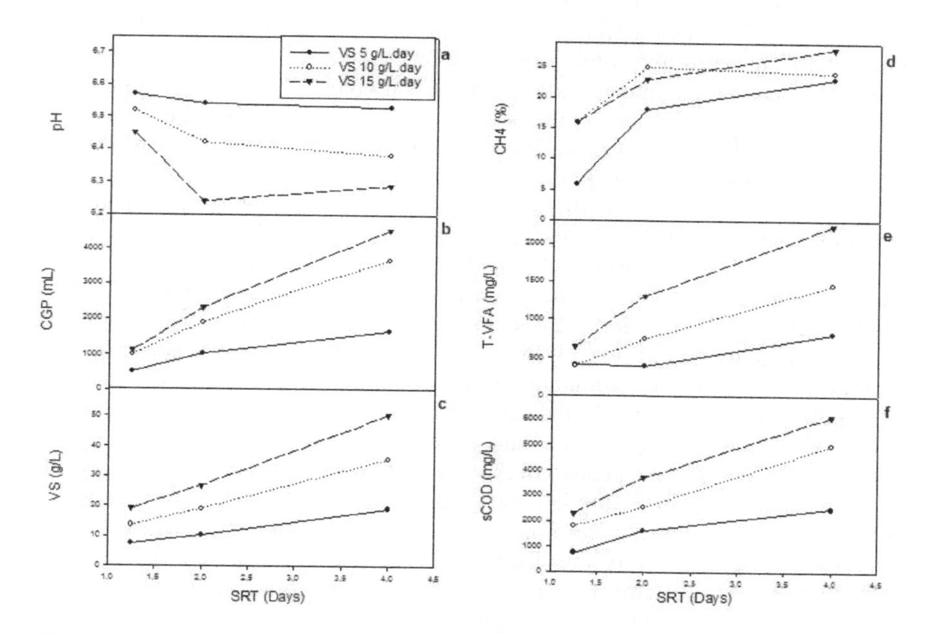

Figure 1: pH, T-VFA, VS, CH_4 percentage, biogas production, and sCOD values observed during the first part of the experiment.

As expected, the increase in the OLR resulted in the increase in the TVFA production (fig. 1b). In addition, the extent of TVFA production for the SRT increase from 1.25 to 2 days was greater than that observed for SRT increase from 2 to 4 days especially for OLRs of 10 and 15 g.VS/L.day. This observation was also verified by the extent of pH drop (being greater for SRT increase from 1.25 to 2 days). These TVFA production trends for all reactors coincided with the sCOD productions (fig. 1c) which increased with the increased OLR and SRT.

The effect of SRT and OLR on TVFA production was also observed for CGP data. As the OLR and SRT increased the CGP in the reactors increased (fig. 1d). It is well known that in addition to VFAs and alcohols both H_2 and CO_2 are produced through acidification. However, GC analyses unexpectedly indicated that methane was produced in all of the reactors studied at varied OLRs and SRTs (fig. 1e). Especially, methane percent of the biogas increased from 5 to

15–27% when SRTs and OLRs were increased to greater values than 1.25 days and 5 g.VS/L.day, respectively, although the pH conditions were close to the optimum operating conditions of highly organic wastes required for acetogenesis. The applied SRT values (1.25 to 4 days) were not favourable for the most sensitive anaerobic bacteria type known as methanogens. The methane production at such low SRTs could be explained by unintentional extended retention times of microorganisms in the reactors due to very high solids concentration and thus lack of homogeneity during daily wasting of sludge. GC analyses also indicated a significant amount of N_2 in the biogas of all reactors changing from 35 to 90% (data was not shown). As expected, denitrication was more dominant at the higher oxidation-reduction potential at the beginning of the experiment. Denitrification might occur during the acidogenic phase, so as to achieve simultaneous VFA production and nitrate elimination, a system could be applied to organic carbon and nitrogen removal from the wastes [2, 28].

Better hydrolysis in acidification process means higher VS reduction. Therefore, in addition to pH and TVFA production, VS is among the critical parameters in determination of the acidification extent of dairy manure known for its high solids content. The average VS concentrations observed in the reactors at varied SRT and OLR combinations were given in fig. 1e. It was observed that increasing the OLR and SRT resulted in the VS accumulation. However, due to the continuous feeding and wasting process, such an accumulation may not clearly indicate the possible VS reduction in the reactors. Therefore, a completely stirred tank reactor (CSTR) system modeling was performed to observe the change in the VS content of the reactors at steady-state conditions. In this CSTR model, each reactor was accepted as reactors which were operated under feeding and wasting process without any destruction/degradation of the feeding. For better comparison, percent VS reduction in each reactor was calculated by considering the theoretical and experimental VS concentrations and given in table 2.

Table 2: The comparison of the reactors with selected parameters.

Reactor	VS Reduction (%)	TVFA (mg/L)	pH
1	8.4	806	6.53
2	14.5	1444	6.38
3	19.5	2236	6.29
4	0	399	6.54
5	8.9	476	6.42
6	14.8	1300	6.24
7	0	412	6.57
8	0	400	6.52
9	2.3	647	6.45

In the first part of the study, the effect of solids and hydraulic retention time (SRT/HRT), organic loading rate (OLR) on the acidification degree was investigated. Results indicated that SRT/HRT and OLR of 2 days and 15 g.VS/L.day, respectively, yielded maximum acidification.

The one-phase conventional configuration (R1) was run as the control for the two-phase configuration (R2) in the second part of the study. All reactors started to produce gas production in the first week of the reactor operation. Gas volumes were measured daily. The results are shown in fig. 2.

The average biogas production values of R1(35), R22(35), R1(25), R22(25) were obtained as 1230±180, 1000±90, 770±70, 290±50 mL/day, respectively. Also, a noteworthy gas production of 130 mL was seen in the mesophilic acidogenic reactor (R21(35)). There were three different gas production trends in fig. 2. This could be explained by the heterogeneous characteristics of the different manure samples collected at different times. This difference resulted in different biodegradability yields. It is clearly seen that temperature affects the performance of the biogas production (fig. 2). The biogas production increased by 60% when the temperature increased from 25°C to 35°C in one-phase reactor. These results are very consistent with literature [18, 29].

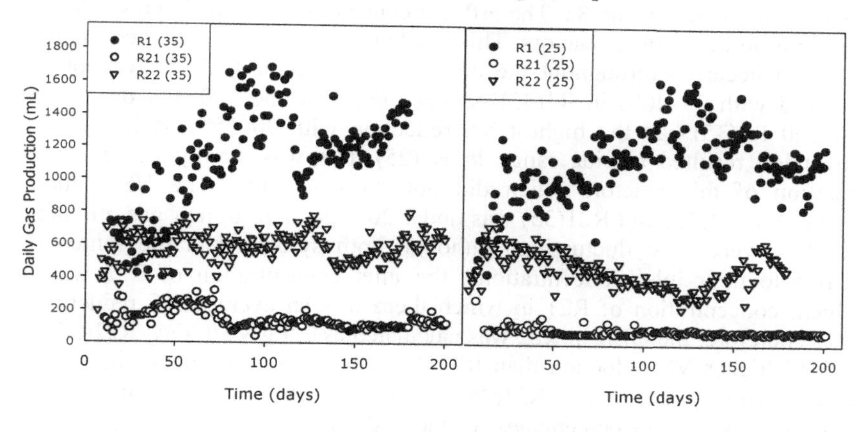

Figure 2: Daily gas productions at 35°C and 25°C.

The average methane content of R1(35), R22(35), R1(25), and R22(25) were determined as 63, 65, 63, and 43%, respectively. The methane yields of these reactors calculated as 221, 216, 132, 43 mL CH_4/g.VS added, respectively. The performances of the reactors in terms of biogas yield could be easily comparable to literature values except R22(25) [1, 18].

When the biogas production yields are compared at mesophilic temperature, the performance of two-phase system (216 mL CH_4/g.VS) is slightly lower than one-phase system (221 mL CH_4/g.VS) in this study. The earlier experiments with fattening-cattle waste had suggested that a HRT of about 20 days was required at 35°C for optimum methanogenic anaerobic digestion and that gas production was reduced significantly at 10 days of SRT [29]. Demirer and Chen [17] demonstrated that a conventional one-phase reactor for unscreened dairy manure at a HRT of 20 days produced 0.235 L biogas/g.VS. When HRT reduced to 10 days, initially an increase was seen in gas production but a few days later an abrupt decline in biogas production were observed, then biogas production was reduced by 90%. In an another work by Wellinger [30], gas yield of straw-rich solid cattle waste was found as 270 and 190 mL/g.VS at HRT of 20 and 10 days, respectively.

From the above discussion, it is obvious that the HRT is directly affecting the biogas production. A simple calculation could reveal which system is preferable in terms of higher biogas production yield. When the HRT of two-phase system is increased from 8.6 to 20 days, the system would produce at least 307 mL CH_4/g.VS instead of 216 mL CH_4/g.VS by using the literature data for the same substrate [30, 31]. Thus, gas production in a two-phase system (R22(35)) would be 42% higher than that of the one-phase system (R1(35)). Moreover, a small amount of produced methane from the acid phase (R21) may also be delivered to R22 or directly collected; it is for sure that methane generation of R2 will also increase.

Volatile solids content is often used as a measure of the biodegradability of the organic fraction of waste. The influent and effluent VS concentrations in the reactors are plotted in fig. 3a. The effluent concentrations revealed a stable trend especially in mesophilic reactors. This stable trend presented that a constant VS reduction occurred throughout the operation. The highest VS conversion was observed with 35–40% in R1(35) between days 10 and 100, but during days 100–200 R2(35) had the highest VS reduction with 30–35%. A 20–30% VS conversion resulted a wide range in R1(25), this was mainly caused by the operation of this reactor, which did not show stability. The VS reduction observed in R2(25) and R21(35) was under 20% parallel to their gas production and they were very fluctuating. Although both systems had the same OLR relative to their inlet concentrations, the inlet concentration of R22 was the effluent concentration of R21 in which there was an average VS reduction of 17%. Therefore, the OLR in R22 was calculated as 2.9 g.VS/L.day. R22(35) had 10–50% higher VS reduction than R22(25), since the performance of R22(25) was low. The VS reduction in R21(25) was nearly below 10% at all times.

As VS conversion percentages, effluent sCOD concentrations had also the same trend (fig. 3b), since the biogas production was due to the degradation of organic compounds. VS and COD parameters could be considered in the same manner as the characteristics of the biodegradability. So, the reduction trends should be similar in terms of VS and COD. The removal of soluble COD concentrations decreased significantly with decreasing temperature. The sCOD reductions of R1(35), R2(35), R1(25) were found as 45, 40, and 55%, respectively. The amount of sCOD in R21(35), R21(25), and R22(25) were increased by 65, 25, and 35%, respectively. The hydrolysis and solubilization of complex materials is the main mechanism in that phase, so that the amount of sCOD increased except R22(25).

The total volatile fatty acids (as HAc) for runs are displayed in fig. 3c. Acetic acid was the dominating VFA in reactors. The effluents of reactors contained mainly acetic acid, propionic and butyric acids, although higher fatty acids were found at lower concentrations.

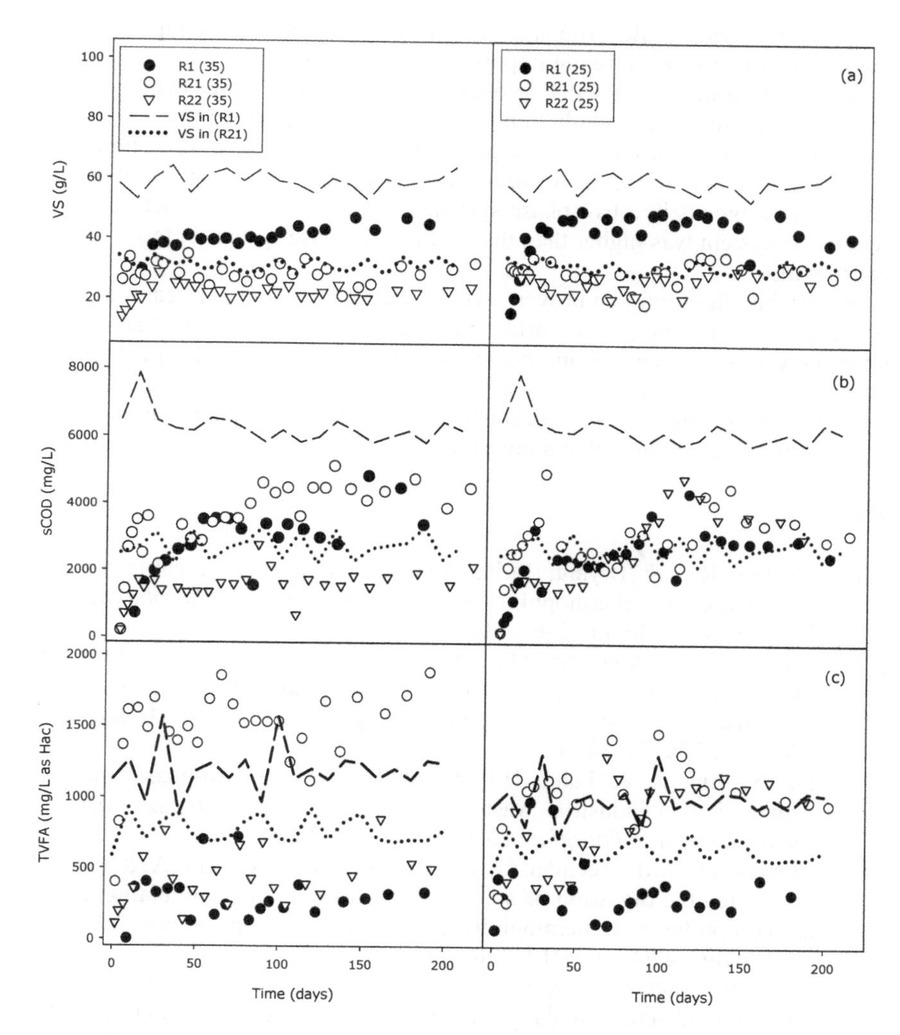

Figure 3: VS, sCOD, and TVFA concentrations in the reactors.

The effluent TVFA concentrations of the first-phase reactor at mesophilic and low temperature operated at 2 day HRT increased to 1700 and 1300 mg/litre (as acetic acid), more than 100 and 60% increase over that of the influent of R21(35) and R21(25), respectively. The effluent VFA concentration of the second-stage reactor in mesophilic temperature decreased to 350 mg/litre (as acetic acid), but the effluent concentration of R22(25) remained the same as expected. The TVFA concentration of R1(25) was much lower than R22(25), since biogas production in R1(25) was more than double of R22(25). This resulted in more VFA consumption in R1(25). Acetic acid was also the predominant VFA in the effluent. The acidogenic efficiency could be increased with temperature, but mixing and pH control were not important parameters [32].

The total effluent VFA value in one-phase reactor was lower than that of the two-phase reactor at mesophilic temperature. It does not mean that more VFAs were converted to methane in one-phase reactor, since more VFA transferred from R21(35) to R22(35). Consequently, higher VFA concentration was converted to biogas in a two-phase system. In other words the efficiency of the two-phase system was higher than that of the one-phase system in terms of VFA consumption.

Anaerobic digestion is a proven technique and at present applied to a variety of waste (water) streams but world wide application is still limited and a large potential energy source is being neglected. Even though several aspects of two-phase AD such as increased stability, lower retention time requirements, liquefaction, etc. are very significant for enhanced AD of manure until now, its application has been limited to a few studies.

References

[1] Mackie, R.I. & Bryant, M.P., Anaerobic digestion of cattle waste at mesophilic and thermophilic temperatures. *Applied Microbiology and Biotechnology*, **43**, pp. 346–350, 1995.

[2] Vigneron, V., Ponthieu, M., Barina, G., Audic, J.M., Duquennoi, C., Mazéas, L., Bernet, N. & Bouchez, T., Nitrate and nitrite injection during municipal solid waste anaerobic biodegradation. *Waste Management*, **27(6)**, pp. 778–791, 2007.

[3] Ke, S., Shi, Z. & Fang, H.H.P., Applications of two-phase anaerobic degradation in industrial wastewater treatment. *International Journal of Environment and Pollution,* **23(1)**, pp. 65–80, 2005.

[4] El-Mashad, H.M., Zeeman, G., Loon, W.K.P. van, Bot, G.P.A. & Lettinga, G., Anaerobic digestion of solid animal waste in an accumulation system at mesophilic and thermophilic conditions, start up. *Water Science & Technology*, **48(4)**, pp. 217–220, 2003.

[5] Lo, K.V., Liao, P.H., & Bulley, N.R., Two-phase mesophilic anaerobic digestion of screened dairy manure using conventional and fixed-film reactors. *Agricultural Wastes*, **17**, pp. 279–291, 1986.

[6] Andrews, J.F. & Pearson, E.A., Kinetics and characteristics of volatile acid production in anaerobic fermentation process. *International Journal of Air and Water Poll*ution, **9**, pp. 439–469, 1965.

[7] Azbar, N. & Speece, R.E., Two-phase, two-stage, and single-stage anaerobic process comparison. *Journal of Environmental Engineering*, **127(3)**, pp. 240–248, 2001.

[8] Elefsiniotis, P. & Oldham, W.K., Effect of HRT on acidogenic digestion of primary sludge. *Journal of Environmental Engineering*, **120**, pp. 645–660, 1994.

[9] Ghosh, S., Ombregt, J.P. & Pipyn, P., Methane production from industrial wastes by two-phase anaerobic digestion. *Water Research*, **19(9)**, pp. 1083–1088, 1985.
[10] Ghosh, S., Buoy, K., Dressel, L., Miller, T., Wilcox, G. & Loos, D., Pilot- and full-scale two-phase anaerobic digestion of municipal sludge. *Water Environment Research*, **67(2)**, pp. 206–214, 1995.
[11] Ince, O., Performance of a two-phase anaerobic digestion system when treating dairy wastewater. *Water Research*, **32(9)**, pp. 2707–2713, 1998.
[12] Keshtkar, A., Meyssami, B., Abolhamd, G., Ghaforian, H. & Asadi, M.K., Mathematical modeling of non-ideal mixing continuous flow reactors for anaerobic digestion of cattle manure. *Bioresource Technology*, **87**, pp. 113–124, 2003.
[13] Nallathambi, G.V., Anaerobic digestion of biomass for methane production: a review. *Biomass and Bioenergy*, **13(1-2)**, pp. 83–114, 1967.
[14] Sung, S. & Santha, H., Performance of temperature-phased anaerobic digestion (TPAD) system treating dairy cattle wastes. *Water Research*, **37**, pp. 1628–1636, 2003.
[15] Ghosh, S., Conrad, J.R. & Klass, D.L., Anaerobic acidogenesis of wastewater sludge. *Journal WPCF*, **47**, pp. 30–45, 1975.
[16] Pohland, F.G. & Ghosh, S., Developments in anaerobic stabilization of organic wastes the two-phase concept. *Environmental Letters*, **1**, pp. 255–66, 1971.
[17] Demirer, G.N. & Chen, S., Two-phase anaerobic digestion of unscreened dairy manure. *Process Biochemistry*, **40(11)**, pp. 3542–3549, 2005.
[18] Varel, V.H., Hashimoto, A.G. & Chen, Y.R., Effect of temperature and retention time on methane production from beef cattle waste. *Applied Environmental Microbiol*ogy, **40(2)**, pp. 217–222, 1980.
[19] APHA (American Public Health Association), Standard methods for the examination of water and wastewater, 19th ed., Washington, DC, 1995.
[20] HACH, HACH Water Analysis Handbook, Loveland, HACH Company, second ed., 1992.
[21] Yilmaz, V. & Demirer, G., Improved anaerobic acidification of unscreened dairy manure. *Environmental Engineering Sc*ience, **25(3)**, pp. 309–317, 2008.
[22] Han, S.K., Shin, S.H., Song, Y.C., Lee, C.Y. & Kim, S.H., Novel anaerobic process for the recovery of methane and compost from food waste. *Water Science and Technology*, **45(10)**, pp. 313–319, 2002.
[23] Kübler, H., & Schertler, C., Three-phase anaerobic digestion of organic wastes. *Water Science and Technology*, **30(12)**, pp. 367–374, 1994.
[24] Verrier, D., Roy, F. & Albagnac, G., Two-phase methanization of solid vegetable waste. *Biological Wastes*, **22**, pp. 163–177, 1987.
[25] Ghosh, S., Improved sludge gasification by two-phase anaerobic digestion. *Journal of Environmental Engineering*, **113(6)**, pp. 1265–1284, 1987.
[26] Wang, J.Y., Xu, H.L., Zhang, H. & Tay, J.H., Semi-continuous anaerobic digestion of food waste using a hybrid anaerobic solid–liquid bioreactor. *Water Science and Technology*, **48(4)**, pp. 169–174, 2003.

[27] Demirer, G.N. & Chen, S., Effect of retention time and organic loading rate on anaerobic acidification and biogasification of dairy manure. *Journal of Chemical Technology and Biotechnology*, **79(12)**, pp. 1381–1387, 2004.

[28] Rustrian, E., Delgenes, J.P., Bernet, N. & Moletta, R., Simultaneous removal of carbon, nitrogen and phosphorus from wastewater by coupling two-step anaerobic digestion with a sequencing batch reactor. *Journal of Chemical Technology and Biotechnology*, **73**, pp. 421–431, 1998.

[29] Summers, R., Hobson, N., Harries, C.R. & Richardson, A.J., Stirred-tank, mesophilic, anaerobic digestion of fattening-cattle wastes and of whole and separated dairy-cattle wastes. *Biological Wastes*, **20(1)**, pp. 43–62, 1987.

[30] Wellinger, A., Process design of agricultural digesters. Nova Energie GmbH Elggerstrasse 36 8356 Ettenhausen, Switzerland. http://homepage2.nifty.com/biogas/cnt/refdoc/whrefdoc/d14prdgn.pdf.

[31] Hobson, P.N. & Wheatley, A.D., *Anaerobic Digestion: Modern Theory and Practice*, Elsevier Applied Science: London, 1993.

[32] Verstraete, W. & Vandevivere, P., New and broader applications of anaerobic digestion. *Environmental Science and Technology*, **28(2)**, pp. 151–173, 1999.

A comparative technology assessment of the anaerobic digestion of an organic fraction of municipal solid waste

A. Cesaro, V. Belgiorno & V. Naddeo
Department of Civil Engineering, University of Salerno, Italy

Abstract

Anaerobic digestion is the biological degradation process of organic matter under anoxic conditions, with production of methane and inorganic by-products, including carbon dioxide. Over the past years, many studies on anaerobic digestion of the organic fraction of municipal solid waste (OFMSW) have been carried out at a laboratory, bench and pilot scale. This work aims to summarize the main features of the OFMSW anaerobic digestion techniques and to define the state-of-the-art of this process in Europe, at both a research and industrial level. To this end, the main experimental results, design solutions and technologies are compared, also in relation to capital and operating costs and data about full scale experiences are reported. Among the different aspects discussed, particular interest has been manifested in pretreatment possibilities in relation to substrate biodegradability and biogas production rate. The data collected and examined in this paper stress the aspects to be improved, in order to define future perspectives of the anaerobic process application for OFMSW treatment.
Keywords: anaerobic digestion, organic fraction of municipal solid waste, pretreatment.

1 Introduction

The need to avoid direct landfilling of biodegradable residues is shared by the whole technical community. In the European Union this statement has been recognized since the promulgation of the Council Directive 1999/31/EC on waste landfilling which, as part of the measures undertaken to improve the sustainability of waste management, forced member states to reduce the amount of biodegradable fractions contained in municipal solid waste (MSW) destined to sanitary landfill [1].

Concerning the above mentioned concept, biological treatments are the main alternatives for the organic fraction of municipal solid waste (OFMSW) [2]. Anaerobic digestion, defined as the biological degradation process of organic matter under anoxic conditions, could be a suitable choice for the biodegradable fraction of urban solid waste. The interest in this treatment is mainly due to the production of methane, which can be used as a renewable energy source in place of aerobic stabilization that requires energy consumption.

The utilization of methane gas as a renewable energy from the biomass anaerobic digestion can be used to obtain certified emission reduction (CER) credits by clean development mechanism (CDM) under the Kyoto Protocol [3]. Over the past years, many studies on anaerobic digestion of OFMSW have been carried out at laboratory, bench and pilot scale. The organic wastes used in previous studies include: market waste [4], fruit and vegetable waste [5], kitchen waste [6], biowaste [7] mechanically sorted [8] and source sorted [9] organic fraction of municipal solid waste. Factors influencing the process stability and efficiency have also been investigated.

This work aims to summarize the main features of the OFMSW anaerobic digestion techniques and to define the state of the art of this process in Europe, at both research and industrial level. To this end, main experimental results, design solutions and technologies are compared, also in relation to capital and operating costs and data about full scale experiences are reported. Among the different aspects discussed, particular interest has been shown in pretreatment possibilities in relation to substrate biodegradability and biogas production rate.

The data collected and examined in this paper stress the aspects to be improved in order to define future perspectives of the anaerobic process application for OFMSW treatment.

2 Pretreatment to anaerobic digestion

Pretreatment is a fundamental step, useful to improve anaerobic digestion yields, operating on the substrate characteristics. Literature data reports experiences dealing with chemical, biological and physical processes (table 1), that are intended to increase the efficiency of the hydrolysis, which is recognized as the rate limiting step in anaerobic digestion processes [10].

Table 1: Pretreatments to anaerobic digestion.

Kind of pretreatment	Objective	Mechanism
Chemical	Improve biodegradability of complex organic matter	Oxidative processes
Physical		Thermal/mechanical processes
Biological		Enzymatic processes

Even though a large number of pretreatments have been investigated, there are very few reports in literature dealing with the possible treatment of the OFMSW prior to anaerobic digestion.

The main chemical processes are the oxidation with wet air and the ozonation, but there are studies dealing with peroxidation methods prior to biosolid anaerobic digestion [11]. Fox and Noike [12] studied wet oxidation for the increase in anaerobic biodegradability of newspaper waste: they found the highest methane conversion efficiency for newspaper pretreated at 190°C, with anaerobic cellulose removals ranging between 74 and 88%.

In latter years, the use of ozone in chemical pretreatment is gaining great interest in the scientific community, not only by removing recalcitrant and toxic compound but also by increasing the biodegradability of waste [10, 13]. Many authors also reported the anaerobic digestion enhancement after an alkaline pretreatment. López Torres and Espinosa Lloréns [10] demonstrated the increased OFMSW anaerobic digestion efficiency after an alkaline pre-treatment with $Ca(OH)_2$.

Since hydrolysis is carried out by extracellular enzymes, some authors investigated the addition of hydrolytic enzymes as a pretreatment to increase the yield and the rate of particulate matter solubilisation during anaerobic digestion. Valladão [14] studied enzymatic pretreatment effects on poultry slaughterhouse effluent anaerobic treatment and he pointed out the enhancement of raw effluent anaerobic treatment efficiency when a 0.1% concentration of enzymatic pool was used in the pre-hydrolysis stage with 1,200 mg/L oil and grease.

Among the physical pretreatments, both mechanical and thermal ones have to be listed. Besides shredding and pulping, which are usually applied to solid waste in order to reduce their particle size, mechanical pretreatment include the use of high pressure gradients to rupture cell walls, but they have been investigated only in relation to sludge, as well as heat treatments [15]. Often thermal pretreatments are combined with alkaline ones. Carrère *et al.* [16] compared thermal and thermo-chemical treatments prior to pig manure anaerobic digestion.

More recent techniques use an ultrasound (US) process to pretreat organic matter to be digested [17]. Chen *et al.* [18] investigated ultrasound process effects on hydrolysis and acidogenesis of solid organic wastes, founding out improved performances.

3 Anaerobic digestion technologies

In relation to the total solid (TS) content fed to the digester, anaerobic digestion (AD) technologies are divided to:

- wet digestion, when the TS < 10%;
- dry digestion, if the TS > 25%
- semi-dry digestion, if the TS content ranges between 10 and 25% [19].

More than 87% of the digestion capacity is provided by single-phase digesters, which can use wet or dry technologies. A slight increase of wet systems was observed as a number of large-scale wet plants were put into operation in Netherland and Spain in 2003 and 2004, while more dry fermentation plants were constructed in 2005. In 2006, De Baere [20] found that dry anaerobic fermentation provides 56% capacity while wet fermentation is used in 44% of the total installed capacity, but according to Schievano *et al.* [21], the wet processes are currently the most widespread.

The list of AD processes and suppliers is highly variable as a result of acquisition, merges and technology advances: the main ones are listed in table 2.

Table 2: Main AD technology suppliers [22].

Supplier	Process	Technology	Capacity range [t/y]
Arrow Ecology	Arrow Bio	wet	90,000–180,000
BTA International GmbH	BTA	wet	1,000–150,000
Citec	Waasa	wet	3,000–230,000
Ros Roca International	Biostab	wet	10,000–150,000
Organic Waste Systems	Dranco	dry	3,000–120,000
Haase	MBT	wet	50,000–200,000
Farmatic Biotech Energy AG	Schwarting – Uhde	wet	18,000–200,000
Valorga International	Valorga	dry	10,000–497,600
Kompogas	Kompogas	dry	5,000–100,000
Strabag (formerly Linde)	Linde – KCA/BRV	wet/dry	6,000–150,000
Entec	Entec	wet	40,000–150,000
Wehlre-Werk AG	Biopercolat	dry	100,000
Global Renewables Ltd	ISKA	dry	88,000–165,000

In the following subsections, data about some of these technologies are given and compared.

3.1 Wet digestion

A wet system appears attractive because of its similarity to the consolidate technology in use for the anaerobic stabilization of sewage sludge coming from wastewater treatment. The operational simplicity is the main reason why this technology is adopted in the majority of plants with capacity lower than 100,000 t/y [22]. In a wet system, the organic solid waste is diluted to less than 10% TS, adding water or recirculating part of the digester effluent. Consequently, CSTR (continuously stirred tank reactor) digesters are mostly used in these applications [19, 23].

Wet processes usually work with low organic loading rate (OLR), ranging between 2 and 4 kg_{VS}/m^3d. Currently, it is still unclear what phenomenon limits the possibility of applying higher organic loads in wet processes. One possible explanation is the concentration of active biomass in the reactor, which could be not high enough. According to other studies, the reason is the nutrient mass short-chain volatile fatty acids. Anyway, possible problems could be easily solved by adding water in order to improve dilution. Table 3 summarizes the typical values of the main single-stage wet anaerobic process parameters.

Table 3: Design parameters and process yields of an AD wet system.

Parameter	Value
Total Solid (TS) content [%]	< 10, until 15
Organic Loading Rate [kg_{VS}/m^3d]	2–4, until 6
Hydraulic Retention Time [d]	10–15, until 30
Process yields	
Biogas production [m^3/t_{waste}]	100–150
Specific biogas production [m^3/kg_{SV}]	0.4–0.5
Biogas production rate [m^3/m^3d]	5–6
Methane content [%]	50–70
Volatile solid (VS) reduction rate [%]	50–60, until 75

Waasa process is implemented in several Finnish plants [23], with operative capacity ranging from 3,000–85,000 t/y. A pulper with three vertical auger mixers is used to shred, homogenize and dilute the wastes in sequential batches. The obtained slurry is then digested in large complete mixed reactors where the solids are kept in suspension by vertical impellers. Gas production ranges between 170 and 320 $Nm^3_{CH4}/t_{VS\text{-}fed}$ and reduction of the volatile solid (VS) feed varies in the range of 40–75% [24].

Wabio technology was supplied in Berlin plant, which treats source sorted wastes with 18–25% TS then diluted till 10–15% TS. The reactor works in mesophilic conditions, with an OLR ranging between 3 and 7 kg_{TVS}/m^3d and a hydraulic retention time (HRT) of 15–17 days. The biogas production ranges between 100 and 150 m^3/t, with methane (CH_4) content varying between 50 and 70%.

Another system quite common in Europe is the Bio-Stab one that was developed by ATU Ingenieurgesellschft für Abfalltechnik und Umweltschutz. This process is a wet digestion technology which makes possible the separation of impurities before the biological treatment by means of mechanical separation without hand-sorting. Due to the very effective separation of the impurities, the digestate has high-quality and is characterized by a low salt and a high organic content. The biogas produced by the digestion is energetically used for the operation of the digestion plant [25].

The BTA process is applicable both in one stage and two stage AD systems. Pilot and industrial scale experiences demonstrate that the BTA process can treat waste with different characteristics, with a moisture content varying between 60% and 90–98%, and a biodegradable matter content ranging between 2% and 50% [26].

In the United Kingdom, one of the established technologies is provided by Monsal Company that exclusively uses wet AD processes. For MSW treatment, their process requires a pre-sorting stage to separate the organic fraction from the bulk of the non-organic materials. Monsal have delivered over 220 anaerobic digestion systems in the last 14 years, supplying digestion technologies to plants ranging from 2,000 to 88,000 m^3 capacity [27]. The company also provides advanced anaerobic digestion systems, which include biological hydrolysis. Bungay [28] discussed three variants of advanced anaerobic digestion using biological hydrolysis and points out its sustainability benefits when compared with more energy intensive thermal hydrolysis processes.

3.2 Dry digestion

During the 80s, research demonstrated that biogas yield and production rate were higher in systems where the waste was kept without diluted water, in its original solid state [29].

In dry systems, the fermenting mass within the reactor is characterised by solids content in the range 20–40%TS: therefore, only dry substrates with TS > 50% need to be diluted. The most widely applied technologies for dry process are Dranco, Valorga and Kompogas, all working in the range 30–40% of TS in the reactor feeding [30].

In the Dranco process, the mixing occurs via recirculation of the waste extracted at the bottom, mixed with fresh wastes (one part fresh waste for six parts digested waste) and pumped to the top of the reactor. This simple design is suitable for the treatment of waste ranging from 20 to 50% TS [23, 31]. Table 4 reports operational parameters and yields of Salzburg (Austria) and Brecht (Belgium) plants implementing the Dranco process.

Table 4: Operational parameters and yield of Salzburg (Austria) and Brecht (Belgium) plants.

Parameter	Salzburg (Austria)	Brecht (Belgium)
Capacity [t/y]	20,000	20,049
Total solid content [%TS]	31	40
Temperature [°C]	55	55
Organic loading rate [kg_{VS}/m^3d]	10	14.9
Specific gas production [m^3/kg_{VS}]	0.36	0.25–0.30
Gas production rate [m^3/m^3d]	4	9.2
Volatile solid removal [%sv]	29	23

The Kompogas process works similarly to the Dranco one, but the plug flow takes place horizontally, in cylindrical reactors. The horizontal plug flow is aided by slowly-rotating impellers inside the reactors, which also serve for homogenization, degassing, and resuspension of heavier particles. This system requires careful adjustment of the solid content around 23% TS inside the reactor. At lower values, heavy particles such as sand and glass tend to sink and accumulate inside the reactor while higher TS values cause excessive resistance to the flow [23, 31].

The digestion plant in Braunschweig-Watenbuttel (Germany), implementing Kompogas process, treats 20,000 t/y of source sorted organic waste. The process yield in terms of biogas production is 80–140 m^3 with 60% of CH_4 per ton of biomass fed while the volatile solid reduction content ranges between 47 and 52% [32].

The biogas plant in Roppen, Tyrol (Austria) was designed to process 10,000 t/y of catering waste (mostly in the touristic peak time, in winter and summer) and kitchen and garden waste from households with the Kompogas technology.

After pretreatment, the waste is fermented in a tube-digester, without the addition of water, at thermophilic temperature (45–60°C). The fermentation end product is separated into a solid and a liquid phase. The former is further stabilized aerobically; the latter is used for moisturising during the aerobic stabilization process of additional organic wastes or optionally used as liquid fertilizer [33].

In Switzerland, Kompogas technology is the most implemented for the anaerobic treatment of organic waste. For some of the oldest plants, operation data are given [34], together with biogas production yields. Table 5 reports the available information and shows that Volketswil plant is overloaded, since the estimated organic loading rate is 17 $kg_{organic\ material}/m^3d$.

Table 5: Kompogas anaerobic digestion plants in Switzerland [34].

Location	Volume [m^3]	Treated waste [t/y]	Gas yields [m^3/y]	Energy [MWh/y]
Bachenbülach	812 (3)	13,577	1,565,361	9,079
Otelfingen	780 (1)	13,814	1,639,904	9,511
Samstagern	512 (2)	9,377	893,944	5,185
Volketswil	290 (1)	~7,500	461,000	2,674

Today, Kompogas has developed a new generation of plants with annual treating capacity ranging from 4,000 to 20,000 t/y. Outside Switzerland, also larger plants have been built. The waste is delivered at ground level inside a hall and the reactor is heated by lignified fraction withheld by the sieve while working up the waste: this procedure increases the usable gas share.

Besides Kompogas, Swiss industrial digestion plants operate with BRV-Linde and Valorga technologies. The BRV-Linde system was implemented in 1994, in the Baar anaerobic digestion plant. It is a thermophilic horizontal plug flow digester, with a working volume of 500 m^3. Since the horizontal cylinder was protected by a patent of Kompogas, the BRV reactor was built with a rectangular cross section. This shape has the disadvantage that it is not possible to stir the content with a longitudinal axis equipped with lateral arms, so it had to be managed with four transversal axes. This system causes a mixing for and backwards in flow direction, hampering the ideal plug flow of the substrate [34].

In the Valorga system, the horizontal plug flow is circular and mixing occurs via biogas injection at high pressure at the bottom of the reactor every

15 minutes [35]. However, differences among dry systems are more significant in terms of sustainable OLR. The Valorga plant at Tilburb (the Netherlands) treats quantities of waste that vary from 400 to 1,100 t/week, in two digesters of 3,300 m^3 each, at 40°C [35]. This corresponds to an OLR of 5 Kg_{VS}/m^3d that is comparable to the design values of plants relying on wet system. An optimized dry system may sustain higher OLR such as the Dranco plant in Brecht (Belgium) that works with OLR of 15 Kg_{VS}/m^3d [20]. This value is achieved without any dilution of the wastes and corresponds to a retention time of 14 days during the summer with 65% VS destruction. Typical design OLR values of the Dranco process are however more conservative (about 12 Kg_{VS}/m^3d) but higher than the wet systems ones. As a consequence, at equal capacity, the reactor volume of a Dranco plant is smaller than that of a wet system [20].

4 Technical and economical comparison

High-solid processes appear to be more efficient at higher loaded process (OLR > 6 kg_{VS}/m^3d) while low solid processes are more beneficial at OLR lower than 6 kg_{VS}/m^3d. Furthermore, wet digestion processes require a relatively high cost for process equipment and the quantity of process water is appreciably greater than that for dry digestion processes: while wet systems typically consume 1 m^3 of fresh water per ton of treated OFMSW, dry systems require water volumes about ten times lower. As a consequence, the volume of wastewater to be discharged is several times smaller for dry systems. For this reason, when wet systems are implemented, it becomes important to have a wastewater treatment plant nearby.

The maximum achievable OLR, however, is highly dependent on reactor configuration. An upper limit on OLR seems to exist at around 15 kg_{VS}/m^3, but the achievable OLR can be greatly affected by the overall digestibility of the waste. With reference to the biogas or methane production rate, comparing the performance of industrial scale OFMSW digesters treating different waste streams is difficult, especially since companies tend to protect performance data. Generalizations have been attempted in the literature: Figure 1 shows the average biogas yield at a given OLR for a large number of laboratory, pilot, and full scale studies [36].

Even though the mesophilic digestion of food waste achieved a biogas yield of about 0.8 m^3/kg_{VS}, the OLR was only 2 kg $_{VS}/m^3d$. For comparison, wet digestion of source sorted organic fraction of municipal solid waste (SS-OFMSW) at 55°C resulted in much lower biogas yields of 0.45 and 0.3 m^3/kg_{VS}, but at OLR of 6 and 9 kg_{VS}/m^3d. Based on this analysis, most of the reactors studied exhibited biogas production rates in the range of 1.5–3.5 m^3/m^3 d.

As regards the economical aspect, according to Confalonieri [37], capital costs range between 400 and 800 €/ton/year of installed capacity. Economy of scale strongly affects capital costs and usually the lower values characterize larger plants (> 50,000–70,000 t/y).

The economical differences between wet and dry systems are small, both in terms of capital and operating costs. The higher costs for waste handling devices such as pumps, screws and valves required for dry systems are balanced by a cheaper pretreatment and reactor, the latter being several times smaller than for wet systems.

Reliable financial information on the performance of the many competing AD technologies is hard to come by; the report of the California Integrated Waste Management Board [36] concluded that discussions of the cost of anaerobic digestion were severely constrained by the lack of real information. Despite the difficulties, data about capital and operating costs were elaborated in two different studies. Although 10 years apart, the capital cost curves from these studies are very similar, while operating costs are not: however, differences could be due to differences in the cost items included in the two works.

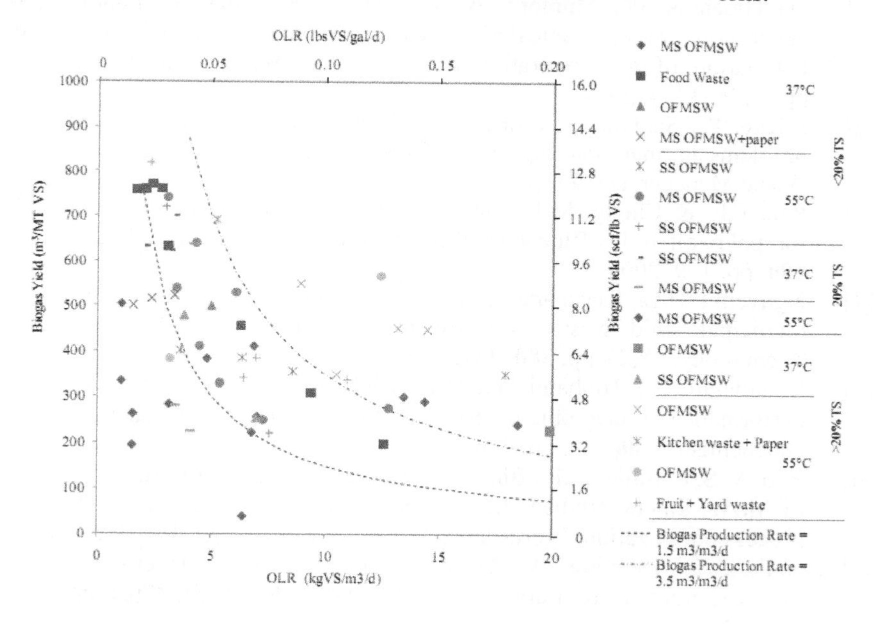

Figure 1: Biogas yields in function of organic loading rate [36].

5 Future perspectives

At present, anaerobic digestion is considered a well-known and reliable technology for OFMSW treatment. The most interesting aspect of this process is the production of biogas, which is a clean and environmentally friendly fuel.

The scientific research is directed towards defining optimisation systems, such as the use of treatments prior to the anaerobic process, in order to enhance its yields, especially in terms of biogas production.

Moreover, research is also carried out on the possibility to recover hydrogen from OFMSW anaerobic digestion in order to use it as energy source, additional to methane.

Finally, an interesting aspect to be studied, especially in relation to wet systems, is the use of reactor supernatants for the growth of particular microalgae used in specific industrial application.

Acknowledgement

The study was partly funded by the University of Salerno (FARB – ex 60%).

References

[1] De Gioannis, G., Muntoni, A., Cappai, G. & Milia, S., Landfill gas generation after mechanical biological treatment of municipal solid waste: Estimation of gas generation rate constants. Waste Management, **29**, pp. 1026–1034, 2009.

[2] Fricke, K., Santen, H., Rainer, W., Hüttner, A. & Dichtl, N., Operating problems in anaerobic digestion plants resulting from nitrogen in MSW. Waste Management, **27**, pp. 30–43, 2007.

[3] Poh, P.E. & Chong, M.F., Development of anaerobic digestion methods for palm oil mill effluent (POME) treatment. Bioresource Technology, **100**, pp. 1–9, 2009.

[4] Nguyen, P.H.L., Kuruparan, P. & Visvanathan, C., Anaerobic digestion of municipal solid waste as a treatment prior to landfill. Bioresource Technology, **98(2)**, pp. 380–387, 2007.

[5] Bouallagui, H., Touhami, Y., BenCheikh, R. & Hamdi, M., Bioreactor performance in anaerobic digestion of fruit and vegetable wastes. Process Biochemistry, **40(3-4)**, pp. 989–995, 2005.

[6] Rao, M.S. & Singh, S.P., Bioenergy conversion studies of organic fraction of MSW: kinetic studies and gas yield-organic loading relationships for process optimization. Bioresource Technology, **95(2)**, pp. 173–185, 2004.

[7] Gallert, C., Henning, A. & Winter, J. Scale-up of anaerobic of the biowaste fraction from domestic wastes. Water Research, **37(6)**, pp. 433–1441, 2003.

[8] Charles, W., Walker, L. & Cord-Ruwisch, R., Effect of pre-aeration and inoculum on the start-up of batch thermophilic anaerobic digestion of municipal solid waste. Bioresource Technology, **100(8)**, pp. 2329–2335, 2009.

[9] Davidsson, Å., Gruvberger, C., Christensen, T.H., Hansen, T.L. & Jansen, J.L.C., Methane yield in the source-sorted organic fraction of municipal solid waste. Waste Management, **27(3)**, pp. 406–414, 2007.

[10] López Torres, M. & Espinosa Lloréns, M.C., Effect of alkaline pretreatment on anaerobic digestion of solid wastes. Waste Management, **28(11)**, pp. 2229–2234, 2008.

[11] Dewil, R., Appels, L., Baeyens, J. & Degrève, J., Peroxidation enhances the biogas production in the anaerobic digestion of biosolids. Journal of Hazardous Materials, **146**, pp. 577–581, 2007.

[12] Fox, M. & Noike, T., Wet oxidation pretreatment for the increase in anaerobic biodegradability of newspaper waste. Bioresource Technology, **91**, pp. 273–281, 2004.

[13] Weemaes, M., Grootaerd, H., Simoens, F. & Verstraete, W., Anaerobic digestion of ozonized biosolids. Water Research, **34(8)**, pp. 2330–2336, 2000.

[14] Valladão, A.B.G., Freire, D.M.G. & Cammarota, M.C., Enzymatic pre-hydrolysis to the anaerobic treatment of effluents from poultry slaughterhouses. International Biodeterioration & Biodegradation, **60(4)**, pp. 219–225, 2007.

[15] Bougrier, C., Delgenès, J.P. & Carrère, H., Impacts of thermal pre-treatments on the semi-continuous anaerobic digestion of waste activated sludge. Biochemical Engineering Journal, **34**, pp. 20–27, 2007.

[16] Carrère, H., Sialve, B. & Bernet, N., Improving pig manure conversion into biogas by thermal and thermo-chemical pretreatments. Bioresource Technology, **100**, pp. 3690–3694, 2009.

[17] Naddeo, V., Cesaro, A., Amodio, V. & Belgiorno, V., Anaerobic co-digestion of municipal solid waste with ultrasound pretreatment. Proc. of the 1st Int. Conf. on Environmental Science and Technology, eds T.D. Lekkas, Global NEST: Chania, Crete, 2009.

[18] Chen, L., Li, B., Li, D., Gan, J. & Kitamura, Y., Ultrasound-assisted hydrolysis and acidogenesis of solid organic wastes in a rotational drum fermentation system. Bioresource Technology, **99**, pp. 8337–8343, 2008.

[19] Hartmann, H., Angelidaki, I. & Ahrin, B.K., Co-digestion of the organic fraction of municipal waste with other waste types. *Biomethanization of the Organic Fraction of Municipal Solid Waste*, ed. J. Mata-Alvarez, IWA Publishing: London, pp. 181–200, 2003.

[20] De Baere, L., Will anaerobic digestion of solid waste survive in the future? *Water, Science and Technology*, **53(8)**, pp. 187–194, 2006.

[21] Schievano, A., D'Imporzano, G., Malagutti, L., Fragali, E., Ruboni, G. & Adani, F., Evaluating inhibition conditions in high-solids anaerobic digestion of organic fraction of municipal solid waste. *Bioresource Technology*, **101(14)**, pp. 5728–5732, 2010.

[22] Foth Infrastructure and Environment, *Source separated organic materials anaerobic digestion*, Feasibility study, 2009.

[23] Lissens, G., Vandevivere, P., De Baere, L., Biey, E.M. & Verstraete, W., Solid waste digestors: process performance and practice for municipal solid waste digestion. *Water Science and Technology*, **44**, pp. 91–102, 2001.

[24] Verma, S., Anaerobic digestion of biodegradable organics in municipal solid wastes, M.Sc. Thesis, Department of Earth & Environmental Engineering, Columbia University, 2002.

[25] Ros Roca International, www.rosroca.com

[26] Bozano Gadolfi, P., La valorizzazione della frazione organica dei rifiuti e delle biomasse con la tecnologia di digestione anaerobica BTA. Tecnologie e prospettive della produzione di energia da biomasse, 2006.

[27] Monsal, www.monsal.com
[28] Bungay, S., Operational experience of advanced anaerobic digestion. *14th European Biosolids and Organic Resources Conference and Exhibition*, Leeds, U.K., 2009.
[29] Oleszkiewicz, J.A. & Poggi-Varaldo, H.M., High solids anaerobic digestion of mixed municipal and industrial wastes. *Journal of Environmental Engineering*, **123**, pp. 1087–1092, 1997.
[30] Bolzonella, D., Pavan, P., Mace, S. & Cecchi, F., Dry anaerobic digestion of differently sorted organic municipal solid waste: a full scale experience. *Water Science and Technology*, **53(8)**, pp. 23–32, 2006.
[31] Vandevivere, P., De Baere, L. & Verstraete, W., Types of anaerobic digester for solid wastes. *Biomethanization of the Organic Fraction of Municipal Solid Waste*, ed. J. Mata-Alvarez, IWA Publishing: London, pp. 111–140, 2003.
[32] Piccinini, S., La digestione anaerobica dei rifiuti organici ed altre biomasse: la situazione e le prospettive in Italia. Il Compostaggio di qualità, Arvan s.r.l. – ISBN 88-87801-08-8, 2003.
[33] Kirchmayr, R., Mayer, M., Braun, R., Krismer, M. & Resch, Ch., Anaerobic digestion of source sorted OFMSW and other co-substrates: status and experience in Austria. Biogas da frazioni organiche di rifiuti solidi urbani in miscela con altri substrati, 2007.
[34] Edelmann, W., Anaerobic digestion of source separated OFMSW and other cosubstrates: status and experience in Switzerland. Biogas da frazioni organiche di rifiuti solidi urbani in miscela con altri substrati, 2007.
[35] Fruteau De Laclos, H., Desbois, S. & Saint-Joly, C., Anaerobic digestion of municipal solid organic waste: Valorga full-scale plant in Tilburg, the Netherlands. *Water Science and Technology*, **36(6-7)**, pp. 457–462, 1997.
[36] California Integrated Waste Management Board, *Current Anaerobic Digestion Technologies Used for Treatment of Municipal Organic Solid Waste*, March 2008.
[37] Confalonieri, A., La digestione anaerobica dei rifiuti urbani in Europa: un'indagine di settore, 2009.

The development of EIA screening for the anaerobic digestion of biowaste projects in Latvia

J. Pubule, M. Rosa & D. Blumberga
Institute of Energy Systems and Environment,
Riga Technical University, Latvia

Abstract

One of the main priorities among European Union member states is to promote renewable energy. Latvia also has to comply with the European Parliament and Council Directive 2009/28/EC on the promotion of renewable energy initiated on 23 April 2009. The directive states that renewable energy must reach 40% of total final energy consumption by 2020. A portion of this can be achieved with the energy produced by burning biogas produced by the anaerobic digestion of biowaste. The use of biowaste as a resource allows Latvia to move closer to the EU's common objectives by reducing the amount of waste disposed in landfills. Biogas production through anaerobic digestion is considered one of the most successful methods of dealing with increasing environmental pollution. Heat and electricity are produced by burning biogas in cogeneration plants. The production of biogas from organic waste also addresses issues related to the disposal of waste products. Waste that would otherwise be considered unusable has now found a second life in the production process of biogas. Given that until 2020, the share of renewable energy in Latvia has to be increased, it can be expected that in the coming years the amount of biogas production and combustion plants will increase. Environmental impact assessment (EIA) is a well-known tool which can aid in sustainable development. Screening is one of the most important stages of EIA. The paper analyses the possibilities of developing a screening process for anaerobic digestion of biowaste projects for EIA in Latvia. In this paper, the demands which should be included in the guidelines of the screening process were created and analysed.
Keywords: EIA, screening, anaerobic digestion, biowaste, biogas projects, renewable energy.

1 Introduction

The aim of the screening phase is to determine if the project shall be subjected to an environmental impact assessment (EIA). Without this verification, some actions would be evaluated very stringently, while others would be forgotten or ignored. While carrying out an effective assessment, a list with the activities planned, accompanied by the values and criteria for determining whether an activity should be evaluated are formed [1, 2].

Categories of activities for which it is compulsorily to carry out an EIA are defined in annex 1 of the EIA directive. In annex 2 of the directive, the operations that need to be undertaken in the EIA procedure on a case by case basis are defined. There may also be activities that are not included in these annexes, but the responsible governmental organization has to decide whether to administer the EIA procedure to the activity after the screening process has been completed. The state may also establish stricter limits, allowing certain projects to be subject to the EIA procedure. The EIA procedure can also be completed on a voluntary basis as an important step in developing the project [3].

Despite the fact that the EIA directive defines a uniform screening process, each country has implemented the EIA directive differently. Together with significant differences in regulations and the practices in the initial inspection, the main shortcomings which have emerged are: doubts of the effectiveness of certain criteria and limits of the systems for projects in annex 1, the non-systematic approach for verification of projects in annex 2, large differences in the initial test criteria between member states, the lack of clear definitions for the types of projects leading to possible misinterpretation of daily practice. Criteria that define the level of significance of the project vary depending on the quantitative or qualitative assessment. Predestined criteria are based on thresholds, or previously taken measurements, and specified restrictions and limits existing in laws, rules, and other guidelines. Criteria based on the judgment are applied, if the project is unlikely to have a significant impact, but in the context the need to take precautions is justified.

Four types of approaches to the initial verification can be distinguished:

1. Pre-trial or preliminary environmental assessment – the need for an EIA is seen through the early assessment process for projects of any type, under any circumstance.
2. Each individual case – the need for an EIA is assessed for each project individually. This approach is usually used in conjunction with another method as its complement, for example, a list of projects or thresholds.
3. A list of projects – the need for an EIA is based on a list of projects divided into different categories and types. There are two types of lists – positive and negative. Positive lists specify the projects that need an EIA, the negative lists show exceptions.
4. Thresholds – the need for an EIA is based on the specific measures and restrictions of pre-defined criteria. These criteria may be differences in the size of the project, the particular location, as well as other criteria [4].

2 Screening in Latvia

In Latvian legislation, activities that must be carried out in an Environmental Impact Assessment are clearly defined. Furthermore, the scope of the work, which is carried out in accordance with international treaties, is clearly defined. At the same time, groups of activities subject to the screening are not well defined in the Latvian legislation. This also holds true in relation to the screening thresholds and criteria for the evaluation of the potential impact of the activity which are also not clearly defined. This leads to a very wide field of subjective judgments and indecisive approaches. By unreasonably reducing the number of projects which are subject to EIA procedure, the public's right to engage in discussion becomes limited, and it is not possible for the public to comment on a project. The lack of laws, regulations, and guidance complicates the decision-making process, leads to a conflict and makes it possible to use the lack of structure within the laws for self-serving purposes.

According to legislation, the application of the initial planned activities is filed at the regional environmental administration of the State Environmental Service. This depends on the location of the planned activity. The regional environmental administration evaluates the material concerning the proposed project and decides whether it is necessary to perform a screening. By evaluating the laws and regulations, it was found that The Cabinet of Ministers of the Republic of Latvia has determined the order in which the screening of the environment is completed for the proposed action. The application procedures and content are defined in the legislation.

According to Latvian laws and regulations on screening, the regional environmental administration is entitled to invite experts to evaluate, request, and receive information from state and local government institutions, as well as to request and receive additional information from the project proposer. However, given the current economic situation, the government is not able to invite experts, due to the lack of resources available for the remuneration of experts [5].

It is not possible to define precisely what is considered to be a significant impact on the environment. In most of the developing world, where the EIA procedure has been introduced, various methodological tools are developed to aid in the process. Consequently, the uniform potential of an environmental impact evaluation in the screening process of proposed actions could be made, and a decision as to whether the activity should include an environmental impact assessment procedure could be taken. In most cases, such aids are made as questionnaires or a matrix.

The criteria of the significance of the impact include the description of the threshold value for identification [6]. The threshold values in Latvia are environmental quality standards, emission limit values, and other limits and restrictions set in various pieces of legislation. Since the various restrictions and environmental quality standards vary in different areas, and for various types of activities, then in most cases the significance of impacts is assessed individually in each case. Often the significance of the impact is not only dependent on the type and amount of hazard of the planned action, but also the characteristics of the selected place have an important role. In some cases, the impacts of small objects which do not exceed the allowable thresholds are potentially dangerous, if they are planned in a sensitive or congested area; therefore, these projects are applied to the EIA procedure. But at the same time, the relatively large objects with possible impact parameters similar to EIA application volumes may not require application of the EIA procedure because of the optimal choice of location, and the projected technology that allows the impacts to be reduced to insignificant levels.

It is clear that screening is one of the most important and responsible steps in the process of the EIA. A faulty decision could lead to substantial financial loss for the future performance of the project, if an unreasonable decision is made to apply the full environmental impact assessment procedure, which requires substantial investments in both time and finances for the project.

Perhaps even greater losses are possible if technical regulations are not fully prepared because the possible impact is not fully assessed for the proposed action, and the implementation of the project has already started, while not realizing the potential problem situations and risk factors resulting in damage to the environment. It is known that in most cases, the consequences of the negative effects requires more resources and time than measures that could have prevented or reduced the possibility of the caused damage.

3 Screening of biogas production projects

Biogas from anaerobic fermentation has several advantages. First of all, biogas is a renewable energy resource. The current global power supply is dependent on fossil fuels; such as crude oil, lignite, coal and natural gas. These are non-renewable energy resources and the reserves of these resources are being depleted much faster than new ones arise. By contrast, the resulting biogas from the anaerobic fermentation process is a completely renewable resource since biogas is produced from biomass, which stores solar energy in the process of photosynthesis. Also, the energy produced by burning biogas contributes to the national energy sustainability and reduces the dependence on imported energy [7].

Similarly, the production of energy by burning biogas is a way to deal with the increasing nature of global warming. Here, the essence lies in the fact that the combustion of biogas releases CO_2, which is the carbon attracted by plants from the atmosphere in the process of photosynthesis. This is the main difference between burning biogas and fossil fuels. In this way, the biogas carbon cycle ends in a very short time.

Biogas from anaerobic fermentation is considered to be an optimal solution for different types of organic waste. The waste is converted into renewable energy and organic fertilizer [7]. Such organic waste as people's household waste, crop residues, animal waste, and fertilizers begins a new life cycle in biogas plants. Otherwise this organic waste would have no use, and new landfills would need to be made for their storage. After anaerobic digestion, the digestate, because of its qualities, serves as a good soil fertilizer. Digestate is rich with nitrogen, phosphorus, potassium and trace elements that can be applied to the soil with conventional liquid manure and slurry equipment. Compared to raw manure, the digestate has improved fertilizer efficiency because it is homogeneous and contains more nutrients; furthermore, the digestate has a better carbon/nitrogen ratio and is nearly odourless.

Biogas production benefits the farmers who have participated, as well as society as a whole. The benefits from the production and usage of biogas are an increased capacity of the local economy, higher employment in rural areas and an improvement in the region's solvency. Compared to fossil fuels, biogas production with anaerobic fermentation requires a much larger work force to ensure the production processes, collection and transportation of anaerobic materials, manufacturing of the equipment, installation, operation, and maintenance of biogas plants.

As for the construction of biogas plants, by the time a project has reached the design stage, a screening must have already been completed. This is in accordance with the law “The Environmental Impact Assessment” annex 2 – “Actions that require a screening”. The application of a screening is filed by the owner of the emerging biogas plant in the State Environment Service regional government, which then completes the procedure in accordance with the law “The Environmental Impact Assessment”.

In order to ensure environmental compliance in the construction project of the biogas plant, the regional environmental administration issues technical regulations, unless it is found that a full EIA procedure should be applied to the biogas plant during the screening period.

4 Waste management in Latvia

Latvia has been divided into 10 waste management regions. In Latvia, waste management is governed by the Waste Management Law. In adherence to the Waste Management Law, waste management in the country has to be achieved according to national and regional waste management plans.

According to 2010 Latvian statistics, 64% of all collected municipal solid waste (and other comparable waste), including bio-degradable solid waste and packaging, was disposed of in MSW landfills [8].

In the next few years, changes in the waste management system are intended to reduce the amount of disposed waste. The Waste Management Law states that waste must be treated before disposal. The treatment of waste before disposal means the separation of recyclable or comestible waste, as well as hazardous waste, by households. In Latvian landfills in the coming years, landfill operators have planned to establish waste pre-treatment lines to meet legislative requirements. That would significantly reduce the amount of waste going to landfills.

An EIA was carried out for 10 out of 11 landfills before the implementation of the MSW landfill project. For one site, an EIA was not carried out since it was made in an existing waste dumping site. In order to evaluate whether predicted impacts were assessed in a sufficiently objective manner, an analysis comparing the current situation with the planned design was carried out. Annual environment reports were used for the performance evaluation of the landfills. Annual environment reports are developed according to the waste management legislation in Latvia.

During the research, an analysis of the biowaste amount, and its characteristics was completed. The results show that the majority of biowaste, which can be used for energy production, is landfilled in Latvia.

From 11 landfills, only 3 have landfill gas collection systems. This leads to the situation where almost all potential waste energy is unused. The amount of landfilled municipal solid waste (MSW) during the five-year period between 2008 and 2012 decreased. This was mostly due to the economic crisis, and a reduction in the number of inhabitants, and not to an increase in the recycling of waste.

From 2008 till 2012, the amount of landfilled municipal solid waste decreased by approximately 30%, from 64,688 tonnes in 2008 to 431,790 tonnes in 2012.

An average of 30% biowaste was used during the calculations. The amount of biowaste which could be separated from municipal solid waste was 219,411 tonnes in 2008 and 151,127 tonnes in 2012.

In fig. 1, the amount of landfilled waste, including municipal solid waste biowaste together with green waste and sludge is shown.

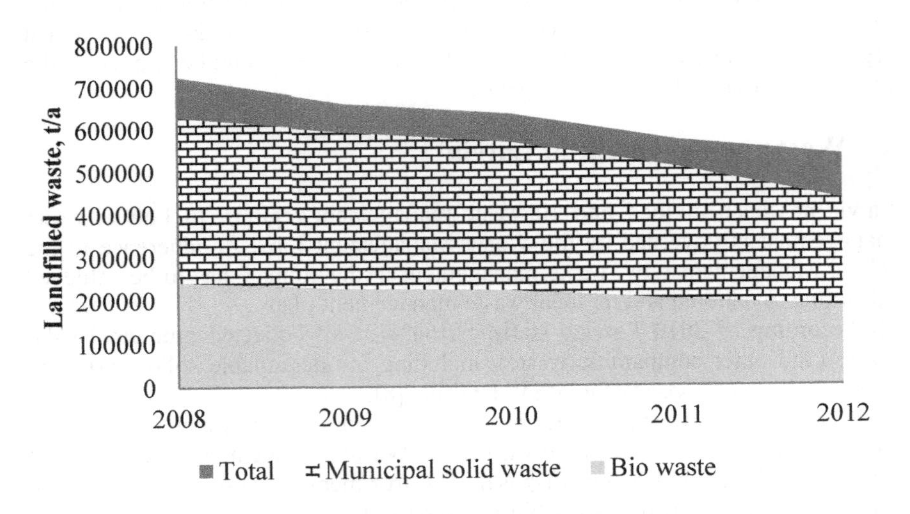

Figure 1: Amount of landfilled waste 2008–2012.

The figure shows a decrease in the amount of landfilled municipal solid waste, but at the same time the amount of landfilled biowaste, which includes green waste, was essentially the same.

5 Biowaste management

Taking into account European targets regarding the reduction of biowaste landfilling, the future consequence will be that suitable biowaste will be used more for the production of energy. In Germany for example, if electricity is generated by facilities using biogas produced by anaerobic digestion of biowaste, this electricity attracts a higher subsidy rate than if the biogas is produced by digesting other types of biomass. Sustainable management of biowaste combines material and energy recovery paths with the aim of optimising the integration of nutrients and carbon recycling, energy production, and CO_2 reduction by replacing fossil fuels.

The treatment option of biowaste depends on the quality of collected materials. If biowaste is separated from the MSW stream, biowaste contains impurities which can negatively affect the operation of the biowaste treatment plant.

Anaerobic digestion of separately collected biowaste leads to a larger energy output compared to the mechanical biological treatment (MBT) of biowaste. Therefore, the separate collection and treatment of biowaste promotes cleaner production principles in biowaste treatment. During the screening process of the two alternatives, anaerobic digestion of separately collected biowaste must be recognised as a more suitable option in comparison with MBT.

Mechanical biological treatment of MSW has become more popular in recent years. MBT is a waste treatment process that involves both mechanical and biological treatment. The first MBT plants were developed with the aim of reducing the environmental impact of landfilling residual waste. The steps of an MBT are shown in fig. 2.

6 Criteria for assessing the impacts of biogas plants

During the EIA of biogas plants, special attention must be paid to these essential criteria: impact on air quality; occurrence of odours; occurrence of noise; impact on soil; impact on water; safety aspects of the station.

6.1 Impact on the air quality

Biogas consists mainly of methane, carbon dioxide, and water vapour. The composition of biogas can be found in table 1.

As shown in table 1, the composition of biogas varies. Mainly, biogas composition depends on the type of substrate that has been used for biogas production. Mainly carbon monoxide (CO), which is a product of incomplete combustion, and nitrogen oxide compounds (NO_x) are emitted while burning biogas. Both of the above-mentioned gases are considered to be greenhouse gases.

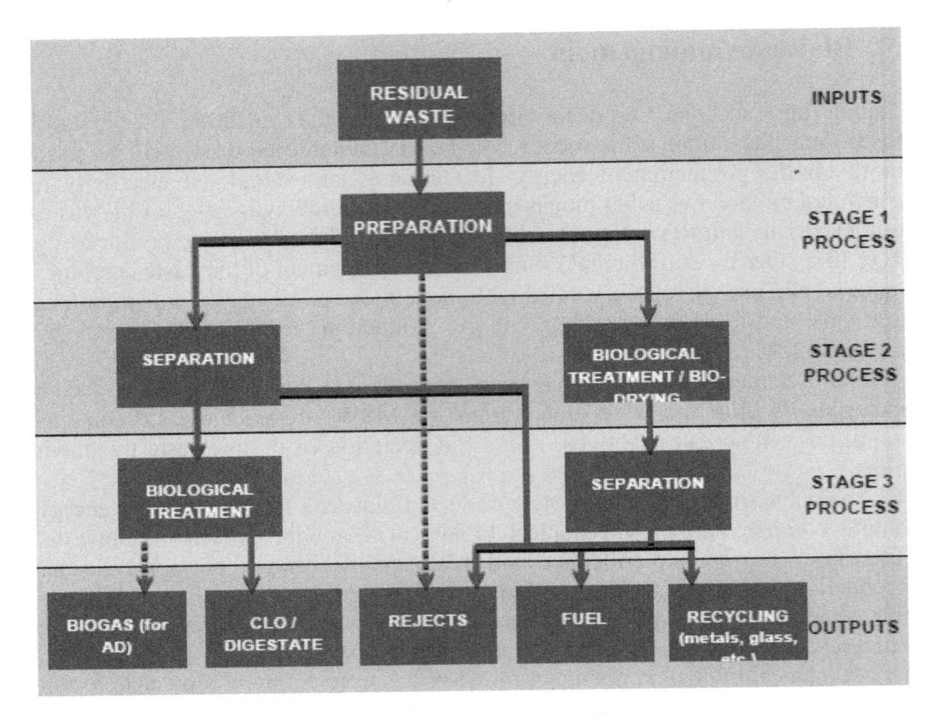

Figure 2: MBT options [9].

Table 1: Biogas composition [10].

Component	Chemical symbol	Content, % by volume
Methane	CH_4	50–75
Carbon dioxide	CO_2	25–45
Water vapour	H_2O	2 (20°C)–7 (40°C)
Oxygen	O_2	<2
Nitrogen	N_2	<2
Other compounds	NH_3	<2

A certain amount of substances are released into the air from the combustion of biogas in the torch. The torch is used when the biogas yield is greater than intended and it is impossible to have additional storage of biogas. Since biogas forms an explosive compound when combined with air, it cannot be simply blown into the atmosphere; instead biogas must be burned in the aforementioned torch. The design of the torch has to be able to convert methane. This reduces incompletely burnt methane and the generation of other products of incomplete oxidation, such as carbon monoxide.

Biogas combustion in the torch is considered to be an environmentally unfriendly solution, because the combustion happens without generating energy and, therefore, should only be regarded as a backup solution. Therefore, special attention must be paid to the solutions pertaining to biogas usage during the initial evaluation of biogas plants.

According to Latvian law, any available and accepted methods in the world can be used for calculating emissions when modelling air emissions. That leads to a situation where different companies use different modelling methods for emissions, as well as different emission factors. As a result, it is not possible to objectively evaluate the estimated emissions from various plants. In addition, companies often use emission factors of natural gas during the modelling process, which should be unacceptable because the biogas composition differs from the composition of natural gas. It is therefore necessary to develop a uniform methodology for calculation and the modelling of air emissions for biogas plants.

6.2 Occurrence of odours

It is necessary that the substrate is kept in closed containers in order to prevent the occurrence of odours. However, odours cannot be completely avoided. This is related to the loading of the substrate into the storage tank. Therefore, an important factor to be considered is the leading wind direction, which certainly should not be directed away from the biogas plant to nearby populated areas. Raw biogas contains hydrogen sulphide. This has a rotten egg smell and is unpleasant for people. So it is necessary to consider all the risks associated with raw biogas emitted into the atmosphere. In contrast, purified biogas is odourless [11].

The smell of manure during anaerobic fermentation is reduced by 80%. Digestate no longer has an unpleasant smell of slurry after treatment – it smells more like ammonia [7].

6.3 Occurrence of noise

Noise emissions during biogas production derive mainly from the production machinery – air and exhaust fans, the mixer cooler, and the flue. Similarly, noise also arises from the transport which delivers the raw material to the biogas plant. If most of the noise from the production installations is unavoidable, then noise arising from transport can be optimized by planning the transportation of raw materials in daylight. In addition, if the resulting noise from the biogas deliveries of raw materials is not permanent, while assuming that the biogas plant operates 24 hours a day, the noise from production installations is a constant. In both cases, how to station the location away from the nearest populated area must be taken into account [7, 10].

6.4 Impact on soil

The end products of the biogas production process are the biogas, which is a fuel gas, and the digestate. Digestate is fermented mass, which is rich in microelements and macronutrients, so it can be used for soil fertilization. The quality of digestate is even better than the untreated slurry. In the case of centralized co-digestion in Latvia, farmers receive only the amount of digestate back which they may use for agricultural fertilizer in accordance with the laws and regulations. The surplus is sold in the particular region to grain growers. In all cases, the digestate is included in fertilizer plants of each farm to replace mineral fertilizers with the digestate.

6.5 Impact on water

It is important that raw materials for biogas production are not present in direct contact with the soil and water. The main reasons for such an event may be insufficient reactor insulation, cracks in storage tanks, and the damage of pipes caused by corrosion. Untreated substrate affects groundwater – the quality drops and pollution increases. Such pollution can lead to various diseases and environmental degradation. Soil and water contamination with raw substrate may cause adverse "slurry vegetation" and increases the risk of spreading pathogens. Related to the issue of quality as well as the protection of groundwater and soil in the life cycle of biogas plants, this occurs in 3 different ways:

1. When the station is at the planning stage; the right combination of place and technology (including materials) can solve most of the issues related to the protection of groundwater;
2. During construction, when a leakage of hazardous substances is possible;
3. In the operating hours of the station, when groundwater monitoring is recommended by taking samples to determine the quality changes.

Also, the risks of flooding in the territory should be taken into account. Biogas plants should be planned in a place that has not been flooded in the past 30 years at least [12].

6.6 Safety aspects of the station

It is necessary to include an assessment of various preventive and damage control measures in the permit required for the biogas plant that allows it to run in the following cases: explosion prevention; fire prevention; mechanical hazards; sound-proof design; electrical safety; lightning protection; thermal safety; asphyxiation and poisoning prevention; hygiene and health safety.

Under certain conditions, biogas combined with air forms an explosive gas mixture.

7 Conclusions

The aim of renewable energy sources can be achieved by 2020 in Latvia. If energy efficiency of user energy resources is ensured, and energy sources are constructed where fossil fuel is replaced with a renewable source.

In the past years, the number of procedures of EIA applied to biogas projects has increased. It is necessary to implement a systematic approach in the procedure of EIA.

It is necessary to elaborate a common approach for the EIA of biogas projects. For establishing and describing the impacts and degree of significance related to carrying out the screening of biogas plants, set criteria ought to be used.

For projects that are subject to the EIA procedure, a life cycle analyses must be done.

References

[1] Toro, J., Requena, I. & Zamorano, M., Environmental impact assessment in Colombia: Critical analysis and proposals for improvement. *Environmental Impact Assessment Review*, **30**, pp. 247–261, 2010.

[2] Kornov, L. & Prapaspongsa, T., Environmental Impact Assessment II, Aalborg University: Aalborg, 2011.

[3] Koornneef, J., Faaij, A. & Turkenburg, W., The screening and scoping of environmental impact assessment and strategic environmental assessment of carbon capture and storage in the Netherlands. *Environmental Impact Assessment Review*, **28**, pp. 392–414, 2008.

[4] Pinho, P., McCallum, S. & Cruz, S., A critical appraisal of EIA screening practice in EU Member States. *Impact Assessment and Project Appraisal*, **28**, pp. 91–107, 2010.

[5] Pubule, J., Blumberga, D. & Romagnoli, F. & Rochas, M., Analysis of the environmental impact assessment of power energy projects in Latvia. *Management of Environmental Quality: an International Journal*, **23(2)**, pp. 190–203, 2012.

[6] Wood, G., Thresholds and criteria for evaluating and communicating impact significance in environmental statements: 'See no evil, hear no evil, speak no evil'? *Environmental Impact Assessment Review*, **28(1)**, pp. 22–38, 2008.

[7] Daublein, D. & Steinhauser, A., Biogas from Waste and Renewable Resources. An Introduction, Second, Revised and Expanded Edition, WILLEY-VCH: Weinheim, 2011.

[8] Ministry of Environmental Protection and regional development, State Waste Management Plan 2013–2020, Ministry of Environmental Protection and regional development: Riga.

[9] Department for Environment, Food & Rural Affairs, Environment Agency. Mechanical Biological Treatment of Municipal Solid Waste. DEFRA: London.

[10] Blumberga, D., Veidenbergs, I., Romagnoli, F. *et al.*, *Bioenerģijas tehnoloģijas*. Madonas Poligrafists: Madona.

[11] Blumberga, D., Dzene, I. *et al.*, *Biogas Handbook*, 2011.

[12] Kossman, W., Pönitz, U., Hebermehl, S. *et al.*, *Biogas Digest, Volume II. Biogas Application and Product Development,* GTZ, 1999.

[illegible] assessing [illegible] challenging [illegible] approach, [illegible] impact [illegible] establishing and [illegible] the impact and degree of significance [illegible] carrying out [illegible] criteria need to be [illegible]

For projects [illegible] life cycle analyses [illegible]

References

[1] [illegible] & Zamorano, M.: Environmental impact assessment [illegible] materials and [illegible] 2016.

[2] [illegible] Environmental Impact Assessment [illegible]

[3] [illegible] The screening and scoping of [illegible] Impact Assessment [illegible] 2015.

[4] [illegible] A critical appraisal of EIA [illegible] Review [illegible]

[5] [illegible] & Lucas, M. [illegible] assessment of power plants [illegible] 2013.

[6] [illegible] environmental impact assessment [illegible] 2016.

[7] [illegible] Revised and expanded edition [illegible] 2011.

[8] [illegible] Environmental Protection [illegible] Ministry of Environmental [illegible]

[9] [illegible] & Rural Affairs [illegible] Treatment of Municipal waste water [illegible]

[10] [illegible] Romagnoli [illegible]

[11] [illegible] 2011.

[12] [illegible] 1998.

High yields of sugars via the non-enzymatic hydrolysis of cellulose

V. Berberi[1], F. Turcotte[1], G. Lantagne[2], M. Chornet[1,3] & J.-M. Lavoie[1]
[1]*Département de Génie Chimique, Université de Sherbrooke, Canada*
[2]*Institut de recherche d'Hydro-Québec (IREQ), Canada*
[3]*CRB Innovations Inc., Canada*

Abstract

Given the cost of cellulosics (quasi-homogeneous residual feeds range in North America, between US$60–80/tonne, dry basis, FOB conversion plant) their fractionation and subsequent use of the intermediate fractions is a strategy that makes economic sense. Furthermore, it permits the isolation of cellulose with low contents of lignin and hemicellulose. Once the cellulose is isolated, its use as a chemically pulped fibre and the conversion of the fines into glucose becomes possible. The authors' group has been working on the chemical depolymerisation of the cellulose (both the fines and the fibres as well) using highly ionic solutions. The method implies recovery of both anions and cations by state of the art technologies. This paper presents the fractionation + ionic decrystallization and depolymerisation approach, provides and discusses its energy balance and compares it with the enzymatic route for hydrolysis in applications to < 40 MML Biofuels/y plants which correspond to < 100 000 t/y of input lignocellulosics, dry basis.
Keywords: cellulose, hydrolysis, depolymerization, electrodialysis, biofuels.

1 Introduction

In North America, residual forest and agricultural biomass cost actually US$60–80 per dry tonne FOB. Such biomass could be considered chemically as quasi-homogeneous since although it may contain the same macromolecules and metabolites (extractives, hemicelluloses, cellulose and lignin), the concentration of each may vary as a function of season, due to weathering and other reasons.

Energy crops or "non-conventional" cultures could also be included in this category since they usually are a "mixture" of different tissues although at this point, this biomass is slightly more expensive, reaching close to US$100 per dry tonne FOB. Conversion of biomass to biofuels and "green chemicals" can be achieved through two general approaches which are categorized as "thermo" or "bio" pathways [1, 2]. Conversion of lignocellulosic biomass could also be achieved through a combination of both [3].

The "bio" approach relies on biological conversion of biomass at one point or the other during the process. This approach is somehow at this point limited by three technological challenges which are 1) the isolation of the cellulose and 2) its hydrolysis to glucose. The last technological aspect (3) that has to be considered is the fermentation of cellulosic sugars which may require additional nutriment to be efficient. The first challenge has been overcome for years by the pulp and paper industry although there is actually a need to develop new techniques leading to the production of pulp which will overall lead to cheaper and less water- and chemicals-consuming processes. Many approaches have been considered to isolate cellulose among which different steam treatments and solvent-related process have been thoroughly investigated [4]. Although isolation of cellulose from the biomass matrix could be performed under different conditions, another key to the economic viability of a biorefinery process is the isolation and utilisation of the other macromolecular fractions of the biomass as lignin and hemicelluloses. Although in many cases, the preliminary conversion process will use thermal and chemical energy, some reports have been made in literature on biological pre-treatment to isolate cellulose [4].

The second key technological challenge that needs to be overcome is the hydrolysis of glucose which is a crucial aspect of the production of cellulosic ethanol. Cellulose is composed of a crystalline and an amorphous phase. In most cases, the amorphous phase is the more vulnerable to hydrolysis, chemical and biological as well. The latter usually relies on a mixture of 3 types of enzymes, endoglucanases, exoglucanases (cellobiohydrolases), and β-glucosidases [5]. The major problems delaying commercialisation of enzyme-based technologies are related on the cost of enzymes. Cellulose is composed of glucose units linked together by acetal bonds and the latter are weakened by an acid catalyst. Therefore, utilisation of acid should in theory be an option, although the major problem in this case is the penetration of the acid in the cellulose crystalline and amorphous structure. Such a concept is not applicable since the cellulose macromolecules are oriented so that the polar functional groups are all linked together via hydrogen bonding making the outside section of cellulose highly hydrophobic. Penetration of water would ease the conversion of cellulose to glucose since it would expose the acetal bonds to any type of Lewis acid. Specific compounds can be used to swell cellulose which means that the swelling molecules can "move" between the microfibrils, reach the cellulosic chains and break its hydrogen bonded structure making a hydrogel. Cellulose is therefore less shelled against an attack since the acetal groups are exposed. The compounds that usually allow such specific interactions are usually ionic. This brings the issue of removing them before fermentation of the cellulosic sugars to avoid inhibition. Another key aspect is to efficiently remove and recover the ions from the mixture.

This paper will discuss the conversion of residual quasi-homogeneous biomass to ethanol. Cellulose has been isolated from the lignocellulosic matrix using the Feedstock Impregnation Rapid and Sequential Steam Treatment (FIRSST) process. The cellulose-rich pulp produced from this process is then hydrolysed to glucose using a non-enzymatic approach involving ionic aqueous solutions. The broth is then purified using a sequential approach with one of the steps being electrodialysis and the remaining sugars fermented to ethanol using industrial grade yeasts. The energy and mass balance of the whole process will be evaluated to see where such a process could be positioned in comparison to other biological techniques leading to the production of cellulosic ethanol.

2 Experimental work

2.1 Feedstock impregnation rapid and sequential steam treatment

Two-step FIRSST process. The scheme of the two-step FIRSST process is depicted in fig. 1 below. After extraction of the secondary metabolites, the biomass was impregnated with water without any catalyst and was then pressed at 6.8 atm (100 psi) to remove the excess water and leave a saturated fiber. After pressing, the biomass is transferred into the 4.5 litres steam gun where about 200 g (dry basis) of chips were cooked at temperatures from 190 to 220°C for 2–5 minutes. Delignification was performed on the pulp obtained from the first steam treatment using a solution of NaOH (2–10%wt of the lignocellulosic material). The wet fibrous solids filtered (as per fig. 1) were then washed again with water 5 times using a water/biomass weight ratio of 5/1. The wet biomass was then impregnated with the alkali solution at 6.8 atm (100 psi) for 5 minutes.

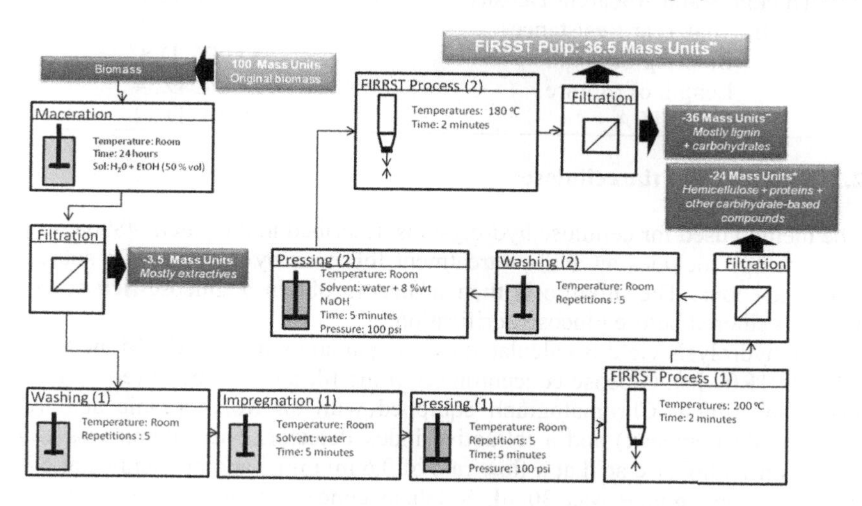

Figure 1: Two steps FIRSST process.

Delignification was performed at a cooking temperature in the 170–190°C range for 2–5 minutes with concentration of NaOH ranging from 2 to 10%wt.

2.1.1 Analysis of the fibres

The testing methods used to evaluate the fibres produced both using FIRSST pulp and kraft pulp (in terms of comparison) is presented in table 1 below.

Table 1: Identification of the test and the standard techniques used for characterization of both FIRSST and kraft pulp.

Test	Standard technique identification
Freeness Determination	ATPPC C.1
Laboratory Screening of Pulp (Pulmac-Type Instrument)	ATPPC C.12
Fibre classification (Bauer-McNett)	ATPPC C.5U
Fibre Length by Automated Optical Analyzer Using Polarized Light	ATPPC B.4P
Forming Handsheets for Physical Tests with Pulp	ATPPC C.4
Forming Handsheets for Optical Tests with Pulp (British Sheet Machine Method)	ATPPC C.5
Grammage	ATPPC D.3
Brightness	ATPPC E.1
Colour Measurement with a Diffuse/Zero Geometry Tristimulus Reflectometer	ATPPC E.5
Opacity	ATPPC E.2
Thickness and Apparent Density	ATPPC D.4
Internal Tear Resistance	ATPPC D.9
Bursting Strength	ATPPC D.8
Length of rupture	ATPPC D.34
TEA	ATPPC D.34

2.2 Hydrolysis of the cellulose

The method used for cellulose hydrolysis is described in the patent #80685-2 [8]. This method includes an acid pretreatment followed by addition of a source of hydroxide ions. The mixture is then heated to obtain a glucose-rich solution, which is filtered before glucose purification.

The hydrolysis yield is calculated by comparison with the ASTM method No E1758-01R07 [9]. Glucose concentration of the filtrate was measured by HPLC, using an Agilent Chromatograph equipped with an RoA-Organic acid (8%) column (Phenomenex) and a refractive index detector. The column was eluted with 5 mM sulphuric acid at a flow rate of 0.6 ml min^{-1} and maintained at 60°C. The injection volume was 30 µl. Sulphate concentration was measured by a colorimetric method [10] and ammonium or sodium concentration, by IC. The apparatus used was a Dionex ICS-3000 ion chromatograph loaded with an IonPac CS12A (2x250mm) and detection was made with electric conductivity.

Three identical cellulosic hydrolyses were performed with 46 g of wet cellulose (64% humidity) each. These tests were done in 2 L erlenmeyer and hydrolysed in an autoclave. The solution was filtered in a Buchner with Glass fiber Fisher Brand filter. The three filtrates were then mixed for the purification step which, in these tests, was made by using electrodialysis only.

2.3 Purification of the cellulosic hydrolysate

A 3 L mixture of the hydrolysis broth (composed of ~300 g/L sodium sulphate, ~300 g/L sulphuric acid and between 10-20 g/L glucose) was purified using an electrodialysis system.

Testing was done at the *Energy technology Laboratory (LTE)* in Shawinigan, QC. The electrodialysis system used was a CS-O batch system from Asahi Glass Co. The cathode and the anode were of iridium oxide and 4 membranes pairs of 180 cm^2 each were used. The experimental conditions were: 30°C, 150–200 L/h and 20 A fixed ($i = 111\ mA/cm^2$). The pH, temperature, conductivity, voltage and intensity were measure automatically at time intervals. Each volume of the three compartments was also measured.

2.4 Fermentation of cellulosic sugars

Inocula were prepared using a medium containing 0.07% w/w gluco-amylase, 16 mM ammonium sulphate, 0.01 g/L Lactrol, 20 g/l yeast (Ethanol Red, Fermentis), 60 ml corn mash (32% solids) and 40 ml water was used. The medium was incubated in an Erlenmeyer flask at 32°C, 150 RPM for 4 hours.

4.6 ml (2 g/l of yeast) of the pre-fermentation medium were then added to 200 ml of the purified lignocellulosic sugars. 8 mM ammonium sulphate and 0.01 g/l Lactrol were also added to the fermentation medium. The corn mash used in the pre-fermentation medium provided the necessary trace nutriments. The medium was incubated in 250 ml Erlenmeyer flasks coupled with a fermentation lock at 34.5°C and 150 RPM during 44 hours.

Monitoring of the fermentation was made by HPLC, using an Agilent Chromatograph equipped with a RoA-Organic acid (8%) column (Phenomenex) and a refractive index detector.

3 Results and discussion

3.1 Feedstock impregnation rapid and sequential steam treatment

The FIRSST process was shown effective for the isolation of the macromolecular structures from different types of lignocellulosic biomass including hardwood (willow), softwood (balsam and fir) and energy crops (hemp and triticale). Production of pulp via the FIRSST process was of 30%wt (dry mass) for willow [6], 40%wt for softwoods [7], 37%wt for hemp and 34%wt for triticale. Pulp produce contained about 3–6% of lignin and the residual fibre was mostly composed of C_6 sugars. Evaluation of the fibres was made both on chemical and mechanical aspects to verify how the severity of the combined steam treatments could be comparable to a classical kraft pulping process. Comparative results are shown in table 2 below.

Table 2: Mechanical properties of FIRSST pulp and kraft pulp for a species of hardwood (*Salix viminalis*) and a mixed species of softwood (*Abies balsamea* and *Picea mariana*).

Test	**H-FIRSST**	**H-Kraft**	**S-FIRSST**	**S-Kraft**
ATPPC C.1 ± 1 mL	409	454	664	721
ATPPC C.12 ± 0.01%	12.7	0.16	5.58	0.11
ATPPC B.4P ± 0.01 mm	0.39	0.41	2.08	2.56
ATPPC D.3 ± 0.1 g/m²	60.9	60.1	59.6	61.0
ATPPC E.1 ± 0.1%	24.7	33.3	24.8	27.6
ATPPC E.5				
L* ± 0.01	64.22	71.45	67.26	69.36
a* ± 0.01	2.82	2.18	4.41	4.12
b* ± 0.01	13.26	12.42	18.31	17.43
ATPPC E.2 ± 0.1%	99.6	99.5	98.5	99.3
ATPPC D.4 ± 0.01 cm³/g	2.17	1.79	2.24	2.36
ATPPC D.9 ± 0.01 mN*m²/g	4.09	3.14	8.04	25.9
ATPPC D.8 ± 0.01 kPa*m²/g	1.35	1.56	3.05	2.28
ATPPC D.34 ± 0.01 km	2.78	4.24	5.03	3.87
ATPPC D.34 ± 0.1 J/m²	15.6	19.6	37.8	22.9

The results of table 2 show that for hardwood and softwood, the two-step steam treatment allowed the isolation of a high quality fibre. These fibres showed less resistance to mechanical stress but showed overall better optical properties. Overall, the FIRSST process allowed the production of fibres that could be converted either to pulp or hydrolysed to glucose depending on market potential.

3.2 Hydrolysis of the cellulose

The cellulosic hydrolysis yields obtained were 96, 83 and 88%, and the average glucose concentration was 10 g/L. These yields can be maximized by changing some parameters, proven by laboratory tests realised to optimise cellulosic hydrolysis.

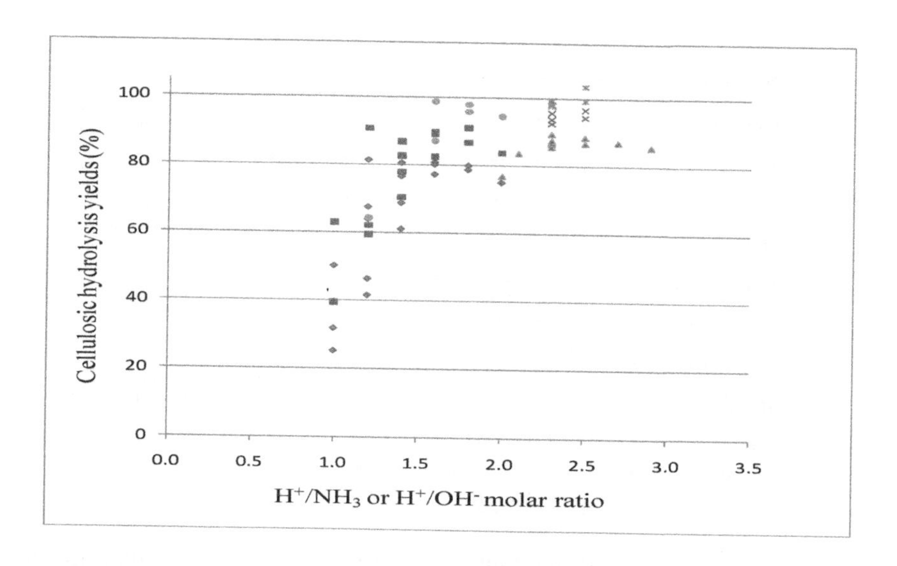

Figure 2: Cellulosic hydrolysis yield versus H^+/OH^- or H^+/NH_3 molar ratio.

3.3 Purification of the cellulosic hydrolysate

Purification of the mixed hydrolysate by electrodialysis leaded to a solution composed of 4.3 g/L sodium sulphate and 1.9 g/L sulphuric acid with glucose. The separation concentrates the glucose solution by a factor of 4.7. With the use of AMV and CMV Selemion membranes (Asahi Glass), it is possible to keep around 95% of the glucose in the diluate. The presence of glucose did not seem to foul the membranes over the time duration of the preliminary process design steps.

Figure 3 presents the composition in sulphuric acid and sodium sulphate of the different compartments versus time:

Due to the voltage increase caused by high ionic concentration in the concentrate (223 g/L sodium sulphate and 126 g/L sulphuric acid)-which means a depletion of the ionic content in the diluate-, the solution in the compartment was replaced by a 50 g/L sodium sulphate solution after 5.7 h of purification.

The energy demand was calculated to 0.57 kWh/kg of separated ions, corresponding to 26 kWh/kg of glucose recovered. Another purification by electrodialysis was realised with a solution containing ammonium sulphate instead of sodium sulphate. This purification had an energy demand of 0.52 kWh/kg of separated ions, corresponding to 24 kWh/kg of glucose. 90% current efficiency was calculated for the test with sodium sulphate and 92% with ammonium sulphate. Since the electrical demand is too high translating into an excessive energy cost per kg of glucose, separation of the ions will have to be done by a combination of less energy intensive steps followed by a final electrodialysis step.

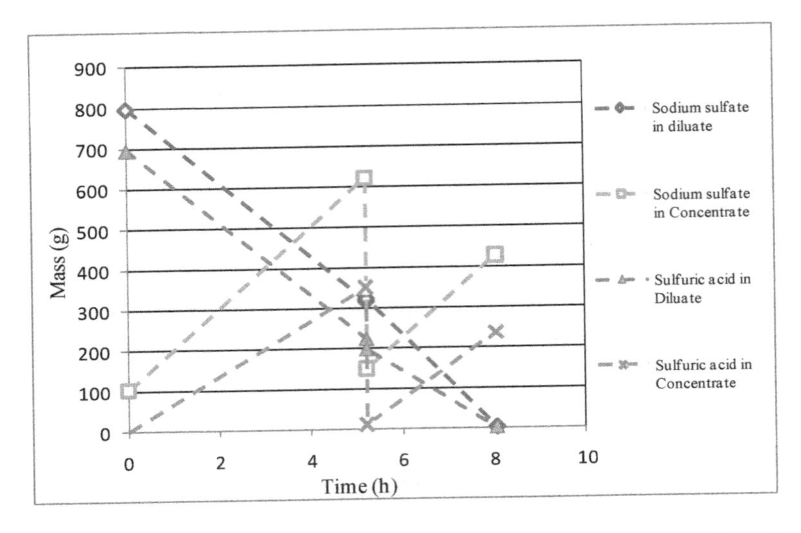

Figure 3: Composition of the diluate and of the concentrate vs. time for the electrodialysis purification.

3.4 Fermentation of cellulosic sugars

2 ml of the fermentation medium were withdrawn after 2, 22 and 44 hours in order to be analysed for ethanol and glucose content, as well and acetic and lactic acid content. The results shown are the mean of duplicates. The progress of the alcoholic fermentation is shown in fig. 4.

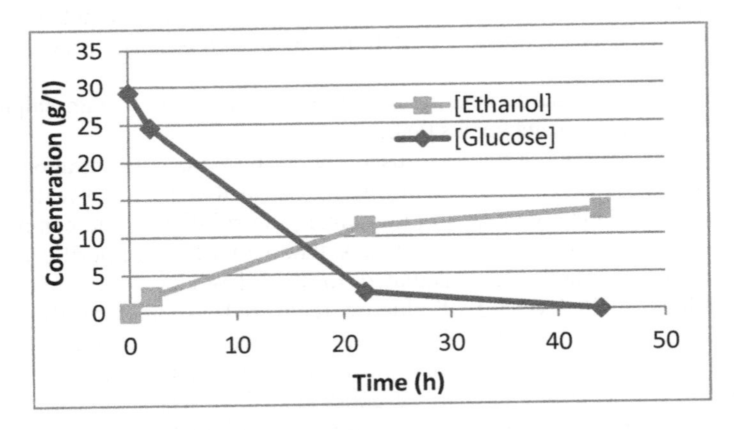

Figure 4: Alcoholic fermentation progress of Ethanol Red *Saccharomyces cerevisiae* yeasts in lignocellulosic sugars medium.

Yields of .46 g/g, an efficiency of 90.1% and a consumption of 100% of the glucose were achieved after 44 hours. No methanol was produced during the fermentation (data not shown). Inhibitors such as acetic acid and lactic acid were also quantified. Table 3 shows the inhibitor composition of the medium through the fermentation.

Table 3: Major fermentation inhibitor concentration through an alcoholic fermentation by Ethanol Red Saccharomyces cerevisiae yeasts.

Time (h)	[lactic acid] mg/l	[acetic acid] mg/l
0	< 100	< 100
2	< 100	< 100
22	< 100	501.0
44	< 100	503.6

Table 3 shows an increase in acetic acid concentration after 22 hours as a by-product of the yeast growth. An acetic acid concentration of .05% (w/v) is considered to have no influence on the yeast growth [11]. The results show that no major inhibitor is in a sufficient concentration to inhibit the yeast fermentation throughout the whole process.

4 Conclusion

The Feedstock Impregnation Rapid and Sequential Steam Treatment (FIRRST) was shown to be suitable not only for the isolation of the cellulosic matrix but also to produce a high quality pulp that could have a potential for paper production. Results have shown that cellulose can be hydrolysed at 90%+ using a strong ionic solution and that this solution could be purified using membrane technologies. Following treatments, the residual glucose solution can easily be fermented with common yeasts to produce cellulosic ethanol. Key for the economics of this approach will be to optimise the energy consumption for the separation process and recovery of ions.

References

[1] Damartzis, T. & Zabaniotou, A., Thermochemical conversion of biomass to second generation biofuels through integrated process design – A review. *Renewable & Sustainable Energy Reviews*, **15(1)**, pp. 366–378, 2011.

[2] Brethauer, S. & Wyman, C.E., Review: Continuous hydrolysis and fermentation for cellulosic ethanol production. *Bioresource Technology*, **101(13)**, pp. 4862–4874, 2010.

[3] Datta, R., Basu, R., Grethlein, H.E., Baker, R.W. & Huang, Y., Ethanol recovery process and apparatus for biological conversion of syngas components to liquid products. PCT Int. Appl. WO 2009108503 A1 20090903, 2009.

[4] Zhu, J.Y., Pan, X. & Zalesny, R.S., Pretreatment of woody biomass for biofuel production: energy efficiency, technologies, and recalcitrance. *Applied Microbiology and Biotechnology*, **87(3)**, pp. 847–857, 2010.

[5] Dashtban, M., Maki, M., Leung, K.T., Mao, C. & Qin, W., Cellulase activities in biomass conversion: measurement methods and comparison. *Critical Reviews in Biotechnology*, **30(4)**, pp. 302–309, 2010.

[6] Lavoie, J.-M., Capek, E., Gauvin, H. & Chornet, E., Production of pulp from Salix viminalis energy crops using the FIRSST process. *Bioresource Technology*, **101(13)**, pp. 4940–4946, 2010.

[7] Lavoie, J.-M., Capek, E., Gauvin, H. & Chornet, E., Production of quality pulp from mixed softwood chips as one of the added value product using the FIRSST process in a general biorefinery concept. *Industrial and Engineering Chemistry Research*, **49(5)**, pp. 2503–2509, 2010.

[8] Chornet, E., Chornet, M. & Lavoie, J.-M., Conversion of cellulosic biomass to sugar. US Provisional Patent Application 80685-2 filed Oct 8, 2008.

[9] ASTM International, Standard test method for determination of carbohydrates in biomass by high performance liquid chromatography. *2008, Annual book of ASTM standards*, pp. 1113–1117, 2007.

[10] HACH, Modèle DR2500, Spectrophotomètre de laboratoire, Procédures, Sulfate, Méthode 8051, pp. 1–7, 2002.

[11] Narendranath, N.V., Thomas, K.C. & Ingledew, W.M., Effects of acetic acid and lactic acid on the growth of *Saccharomyces cerevisiae* in a minimal medium. *Journal of Industrial Microbiology and Biotechnology*, **26(3)**, pp. 171–177, 2001.

Sunflower biodiesel: efficiency and emissions

J. A. Ali[1] & A. Abuhabaya[2]
[1]*Koya University, Kurdistan*
[2]*University of Huddersfield, UK*

Abstract

With economic development, energy needs grew, utilizing natural resources such as wood, fossil fuels, and nuclear energy in the preceding century. However, rising concerns on energy security and climate change in recent years have focused attention on using alternative sources of energy such as bio-fuels. Bio-fuels are produced from renewable resources, particularly plant-derived materials, and its production provides an alternative nonfossil fuel without the need to redesign current engine technology. This study presents an experimental investigation into the effects of using bio-diesel blends on diesel engine performance and its emissions. The bio-diesel fuels were produced from sunflower oil using the transesterification process with low molecular weight alcohols and sodium hydroxide then tested on a steady state engine test rig using a four cylinder Compression Ignition (CI) engine. The paper also shows how by blending bio-diesel with diesel fuel harmful gas emissions can be reduced while maintaining similar performance output and efficiency. Production optimization was achieved by changing the variables which included methanol/oil molar ratio, NaOH catalyst concentration, reaction time, reaction temperature, and rate of mixing to maximize bio-diesel yield. In addition, a second-order model was developed to predict the bio-diesel yield if the production criteria is known. It was determined that the catalyst concentration and molar ratio of methanol to sunflower oil were the most influential variables affecting percentage conversion to fuel.
Keywords: bio-diesel, transesterification, optimization, sunflower oil, engine performance and emission.

1 Introduction

Energy is very important for humans as it is used to sustain and improve their well-being. It exists in various forms, from many different sources. The concerns over global warming and energy security have raised the issue of using alternative sources of energy such as bio-fuels produced from renewable resources such as plant. There are mainly two types of bio-fuels (first generation bio-fuels): ethanol – produced by fermentation of starch or sugar (such as grains, sugarcane, sugar-beet) and bio-diesel – produced by processing vegetable oils (such as sunflower, rapeseed, palmoil). Another type of bio-fuel is cellulosic ethanol known as second generation bio-fuel, produced mainly from wood, grasses and other lignocellulosic materials from renewable sources.

Bio-fuels have become a high priority in the European Union, Brazil, the United States and many other countries, due to concerns about oil dependence and interest in reducing greenhouse gas emissions. The European Union Bio-fuels Directive required that member states realize a 10% share of biofuels (on energy basis) in the liquid fuels market by 2020 [1]. For bio-diesel production, most of the European countries use rapeseed and sunflower oil as their main feedstock while soybean oil is the main feedstock in the United States. Palm oil in South-east Asia (Malaysia and Indonesia) and coconut oil in the Philippines are being considered. In addition, some species of plants yielding non-edible oils, e.g. jatropha, karanji and pongamia, may play a significant role in providing resources.

Bio-diesel is derived from vegetable oils or animal fats through transesterification [2] which uses alcohols in the presence of a catalyst that chemically breaks the molecules of triglycerides into alkyl esters as bio-diesel fuels with glycerol as a by-product. The commonly used alcohols for the transesterification include methanol and ethanol. Methanol is adopted most frequently, due to its low cost.

Engine performance testing of bio-diesels and their blends is indispensible for evaluating their relevant properties. Several research groups have investigated the properties of a bio-diesel blend with soybean oil methyl esters in diesel engines and found that CO and soot mass emissions decreased, while NOx increased. Labeckas and Slavinskas [3], examined the performance and exhaust emissions of rapeseed oil methyl esters in direct injection diesel engines, and found that there were lower emissions of CO, CO_2 and HC. Similar results were reported by Kalligeros *et al.* [4], for methyl esters of sunflower oil and olive oil when they were blended with marine diesel and tested in a stationary diesel engine. Raheman and Phadatare [5], studied the fuel properties of karanja methyl esters blended with diesel from 20% to 80% by volume. It was found that B20 (a blend of 20% bio-diesel and 80% petroleum diesel) and B40 could be used as an appropriate alternative fuel to petroleum diesels because they apparently produced less CO, NOx emissions, and smoke density. A technique to produce biodiesel from crude Jatropha curcas seed oil (CJCO) having high free fatty acids (15%FFA) has been developed by Berchmans and Hirata [6]; a two-stage transesterification process was selected to improve the methyl ester yield.

Lin *et al.* [7] confirmed that emission of polycyclic aromatic hydrocarbons (PAH) decreased when the ratio of palm bio-diesel increased in a blend with petroleum diesel. In general, bio-diesel demonstrated improved emissions by reducing CO, CO_2, HC and PAH emissions though, in some cases, NOx increased.

The objective of this study was to optimize the production of bio-diesel from sunflower oil within a laboratory environment and to evaluate its effectiveness through testing using a laboratory engine test rig. The results showed improved engine performance and reduced exhaust gas emissions with levels acceptable to the standard ASTM D6751 (which was correlated to the content of pigments such as gossypol) [8]. A literature search indicated that little research has been conducted using RSM to analyse the optimal production of biodiesel using vegetable oils. This study intended to make use of the RSM process to maximize the production of bio-diesel from sunflower oil using the conventional transesterification method. In addition to using the RSM for optimizing the methanolysis of sunflower oil, it was also desired to develop a mathematical model which would describe the relationships between the variables and so allow yield to be predicted before the production process was finalised.

2 Materials and methods

2.1 Materials

The materials used in this study including methanol, sodium hydroxide and sunflower oil were purchased from Fisher Scientific and local shops in the city of Huddersfield, United Kingdom. The bio-diesel from sunflower oil was blended at B5 (5% of bio-diesel to 95% of standard diesel by volume), B10, B15 and B20 and evaluated for engine performance and exhaust gas emissions compared to standard diesel.

2.2 Equipment

Experiments were conducted in a laboratory-scale setup. A 500 ml, three-necked flask equipped with a condenser, a magnetic stirrer and a thermometer was used for the reaction. The flask was kept in the 35 °C water bath and stirring speed was maintained at 200 rpm. The reaction production was allowed to settle before removing the glycerol layer from the bottom, and using a separating funnel to obtain the ester layer on the top, separated as bio-diesel.

2.3 Fatty acid profile

In accordance with the approved method of the American Oil Chemists Society (AOCS), eqn (1) was used to calculate the FFA content of vegetable oils:

$$\%\text{Free Fatty Acid (as olieic acid) } FFA = \frac{Tv \times M \times 28.2}{W} \quad (1)$$

where *Tv* is the titration value (ml of NaOH), *M* the Molarity of NaOH (0.025M), and *W* the mass of oil sample (g).

3 Experimental setup

3.1 Biodiesel production process

The presence of NaOH is meant to produce methyl esters of fatty acids (biodiesel) and glycerol as shown in fig. 1. In this study, the reaction temperature was kept constant, at 35 °C. The amount of methanol needed was determined by the methanol/oil molar ratio. An appropriate amount of catalyst dissolved in the methanol was added to the precisely prepared sunflower oil. The percentage of the biodiesel yield was determined by comparing the weight of up layer bio-diesel with the weight of sunflower oil added. Figure 1 shows the reaction conversion of vegetable oils to bio-diesel.

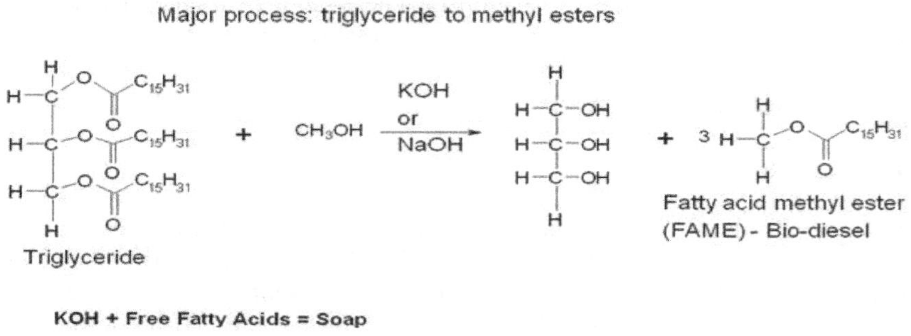

Major process: triglyceride to methyl esters

Triglyceride + CH_3OH $\xrightarrow[\text{NaOH}]{\text{KOH or}}$ Glycerol + 3 Fatty acid methyl ester (FAME) - Bio-diesel

KOH + Free Fatty Acids = Soap

Figure 1: Chemical reaction for sunflower bio-diesel production.

3.2 Engine test setup

The performance of the bio-diesel produced by the transesterification process was evaluated on a Euro 4 diesel engine mounted on a steady state engine test bed. The engine was a four-stroke, direct injection diesel engine, turbocharged diesel, 2009 2.2L Ford Puma Engine as used on the range of Ford Transit vans. The general specification was Bore = 89.9 mm, stroke = 94.6 mm, engine capacity = 2402 cc, compression ratio = 17.5:1, fuel injection release pressure = 135 bar, max power = 130 kW @ 3500 rpm, max torque = 375.0 Nm @ 2000–2250. Emissions were measured using a Horiba EXSA 1500 system, measuring CO_2, CO, NOx and THC. The test procedure was to run the engine at 25, 50, 75 and 100% engine load over a range of predetermined speeds, 1500, 2200, 2600, 3000 and 3300 rpm. At each of these settings, the torque, fuel consumption and emissions were measured, the standard diesel forming the benchmark.

3.3 Trials and optimization

Optimization of the transesterification process was conducted via a 3-factor experiment to examine effects of methanol/oil molar ratio (*M*), reaction time (*T*), and catalyst concentration (*C*) on yield of methyl ester using a central composite rotatable design (CCRD). The CCRD consisted of 20 experimental runs ($2k + 2k + m$, where k is the number of factors and m the number of replicated centre points), eight factorial points ($2k$), six axial points ($2 x k$), and six replicated centre points ($m = 6$). Here k is the number of independent variables, and $k=3$ should provide sufficient information to allow a full second-order polynomial model. The axial point would have $\alpha = 1.68$. Results from previous research [9] were used to establish a centre point of the CCRD for each factor. The centre point is the median of the range of values used: 6/1 for methanol/oil molar ratio, 1% catalyst concentration and 60 min reaction time. To avoid bias, the 20 experimental runs were performed in random order. Design-Expert 8.0 software was used for regression and graphical analyses of the data obtained. The experimental data was analyzed using response surface regression (RSREG) procedure in the statistic analysis system (SAS) that fits a full second-order polynomial model

$$y = \beta_0 + \sum_{i=1}^{3} \beta_i x_i + \sum_{i=1}^{3} \beta_{ii} x_i^2 + \sum_{i=1}^{3} \sum_{j=1}^{2} \beta_{ij} x_i x_j \qquad (2)$$

where y is % methyl ester yield, x_i and x_j are the independent study factors, and β_0, β_i, β_{ii} and β_{ij} are intercept, linear, quadratic, and interaction constant coefficients, respectively. A confidence level of $\alpha = 5\%$ was used to examine the statistical significance of the fitted polynomial model.

The RSREG procedure uses canonical analysis to estimate stationary values for each factor. Using the fitted model, response surface contour plots were constructed for each pair of factors being studied while holding the third factor constant at its estimated stationary point. Confirmatory experiments were carried out to validate the model using combinations of independent variables that were not a part of the original experimental design but within the experimental region.

4 Results and discussion

4.1 Fatty acid content analysis

Since higher amounts of free fatty acid (FFA) (>1%w/w) in the feedstock can directly react with the alkaline catalyst to form soaps, which can then form stable emulsions and prevent separation of the bio-diesel from the glycerol fraction and decrease the yield, it is better to select reactant oils with low FFA content or to reduce FFA in the oil to an acceptable level before the reaction. Nevertheless, the FFA (calculated as oleic acid) content of the sunflower oil used in this experiment was, on average, only 0.13% which was within acceptable levels to be directly used for reaction with the alkaline catalyst to produce bio-diesel [10].

The remaining main factors affecting the transesterification include reaction time, temperature, molar ratio, rate of mixing, and catalyst concentration.

4.2 Modelling the biodiesel production

The regression coefficients for a second-order polynomial model have been determined from Design-Expert 8.0 software program; using these coefficients, the predicted model in terms of uncoded factors for methyl ester yield is given by

$$Y_{yield} = -121.52 - 1.29T + 32.05M + 183.66C + 0.49TM - 0.59TC - 4.44MC - 0.05T^2 - 1.99M^2 - 62.91C^2 \quad (3)$$

where Y_{yield} is the methyl ester yield, and *T*, *M* and *C* are the actual values of the test variables. These coefficients show that the linear terms for methanol/oil molar ratio and catalyst concentration (*M* and *C*, respectively), the quadratic terms in M^2 and C^2, and the interaction terms in *TC* and *TM* had significant effects on the yield. Among these, *M*, *C*, C^2 and *MC* were significant at the significance level, while M^2 and *TM* were significant at the level.

The results suggest that linear effects of changes in molar ratio (*M*) and catalyst concentration (*C*) and the quadratic effect C^2 were primary determining factors on the methyl ester yield as these had the largest coefficients. That the quadratic effect, M^2 and the interaction effect *MC* were secondary determining factors and those other terms of the model showed no significant effect on Y_{yield}. Positive coefficients, as with *M* and *C*, enhance the yield. However, all the other terms had negative coefficients. Figure 2 show the predicted yield obtained by eqn (3) denoted by solid line and comparison with experimental measured yield for twenty runs.

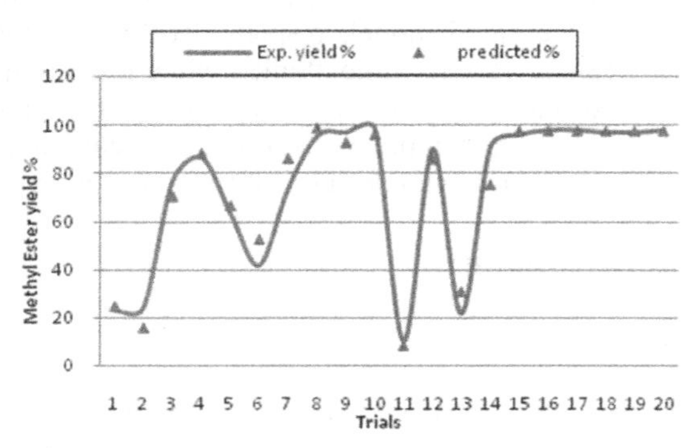

Figure 2: Biodiesel production predicted data and comparison with experimental measured data.

4.3 Engine performance and emission results

4.3.1 Brake performance analysis

Due to the lower calorific value of biodiesel, both torque and brake power reduces. Figures 3 and 4 show the effect of standard diesel and biodiesel fuel on brake power and torque respectively. However, the differences between standard diesel and biodiesel were very small in most cases. Figure 5 presents the effects of standard diesel and biodiesel fuel on BSFC; a bigger difference is shown in higher speed engine. The increase of BSFC may be attributed to the higher density, higher fuel consumption and lower brake power due to lower calorific value of the biodiesel.

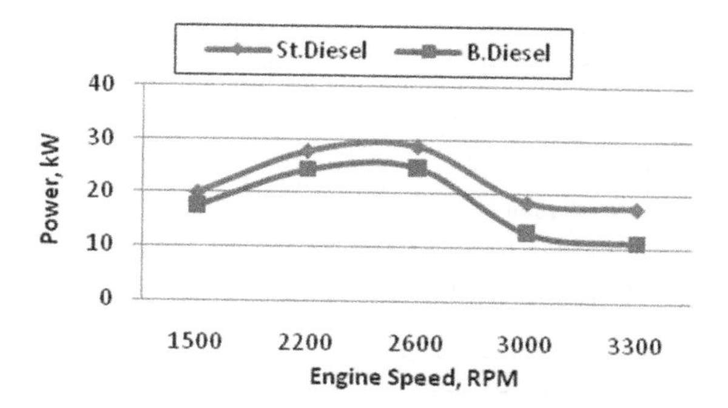

Figure 3: Variation of the brake power with the engine speed, at full load.

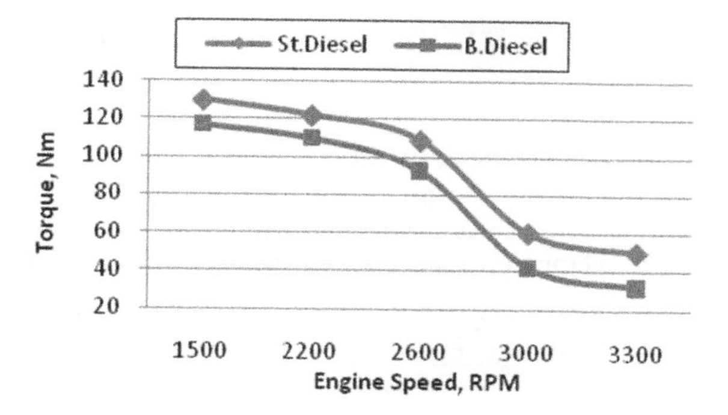

Figure 4: Variation of the brake torque with the engine speed, at full load.

4.3.2 Engine performance analysis

Sunflower oil itself has relatively low energy content, but the bio-diesel fuel produced from it has a value of about 37.5 MJ/kg, close to that of petroleum diesel; this means that efficiency and output is lower but only by a small percentage. Figures 6 and 7 show the curves for engine power and torque respectively. By simple proportions the energy content of the blend can be calculated. Energy content of blend = (%diesel × 42.5 + %bio-diesel × 37.5). It can be seen from fig. 6 that the loss in power is close to the value predicted. At 20% bio-diesel the calculated power is 41.5 MJ/kg, a decrease of 2.35% compared to petroleum diesel, the measured decrease was about 1.72%.

The same trend in the results was seen for torque, there was a progressive decrease in torque as the proportion of bio-diesel in the blend increased (see figs. 6 and 7). The decrease in torque was more apparent than that of the power, because diesel engines are more focused on torque curves than power curves.

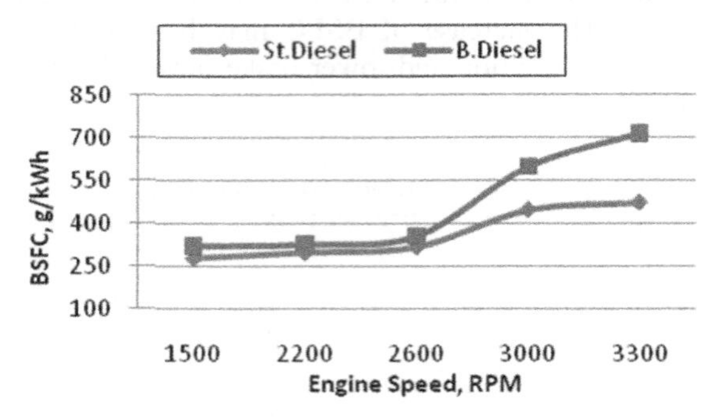

Figure 5: Variation of BSFC with the engine speed, at full load.

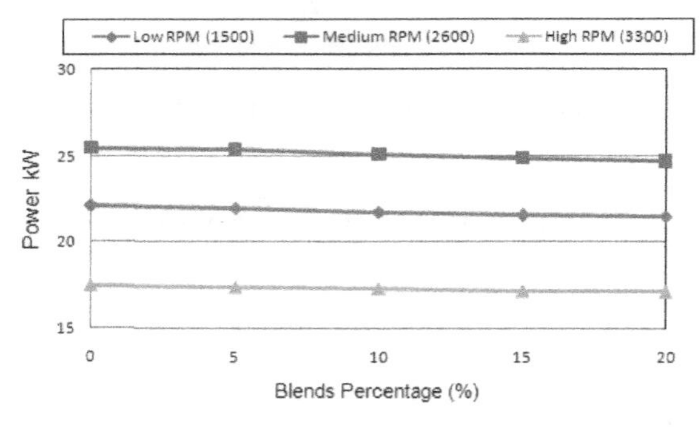

Figure 6: Average power output for different bio-diesel blends.

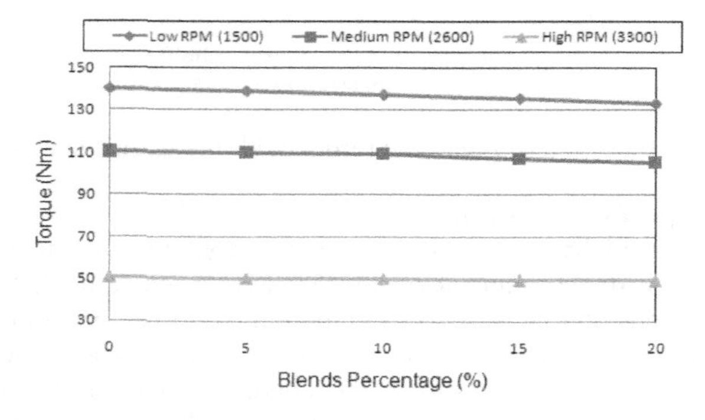

Figure 7: Torque output for different bio-diesel blends.

4.3.3 Engine exhaust gas emissions analysis

As was stated previously the results of bio-diesel blend fuels over the petroleum diesel should show decrease in the emissions of CO, HC, with a slight increase in NOx, and overall similar values for CO_2. This trend can be seen in fig. 8.

When bio-diesel is present there is additional carbon, hydrogen and oxygen to be added to the reaction. The resulting problem is seen at B5, this additional carbon caused the emitted CO_2% to increase. This then falls as the proportion of bio-diesel is increased and a state similar to that for diesel fuel is reached at about B20. Following this trend it is estimated that at higher concentrations of bio-diesel blends (> B20) the CO_2% emitted would actually be lower than for diesel fuel.

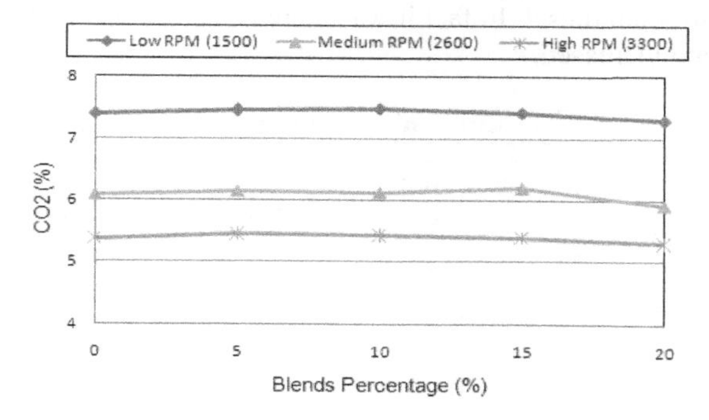

Figure 8: Variation of carbon dioxide emissions for different biodiesel.

The second emission to be analyzed is CO. Carbon Monoxide is present when dissociation is present in the combustion due to incomplete combustion. Figure 9 shows the CO emission for the bio-diesel obtained from sunflower oil. From the data it was clear that the CO emission decreased as the bio-diesel blend increased.

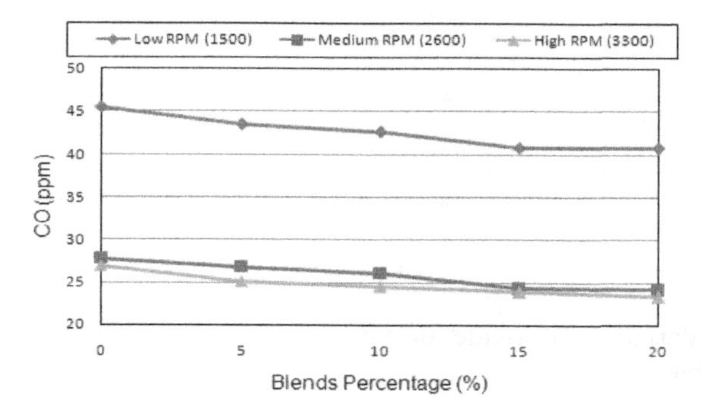

Figure 9: Variation of carbon monoxide emissions for different biodiesel blends.

Hydrocarbon emission should be reduced by the use of bio-diesel. From the data in fig. 10 there was significant and substantial decrease in HC emissions. As the combustion becomes more complete less dissociation occurs yielding fewer hydrocarbons in the emissions. The decrease in HCs from over 40 ppm to less than 30 ppm is good for a fuel which is as efficient as diesel fuel but friendlier to the environment. An oxide of nitrogen (NOx) was the only emission which did not seem to show a decrease relative to diesel fuel. In fact it was increasing steadily as the percentage of bio-diesel blend increased, see fig. 11. From the data it was apparent that the change is only being incremented at B20 by a maximum value of 3.21%, yet with a mean more resembling that of 2.33%. An oxide of nitrogen is the only emission which did not seem to show a decrease relative to petroleum diesel. In fact it was increasing steadily as the percentage of bio-diesel blend increased.

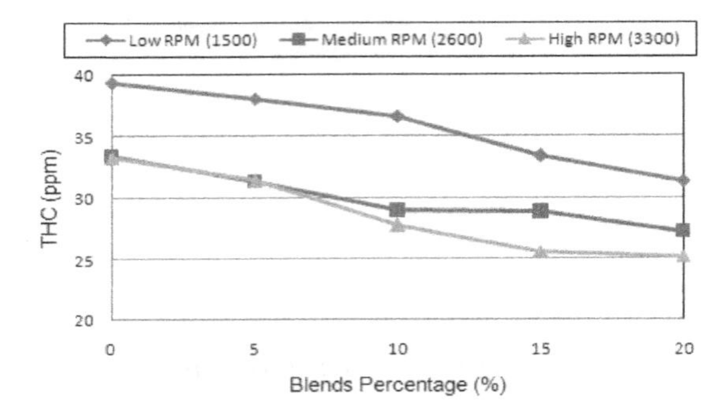

Figure 10: Variation of hydrocarbon emissions for different biodiesel blends.

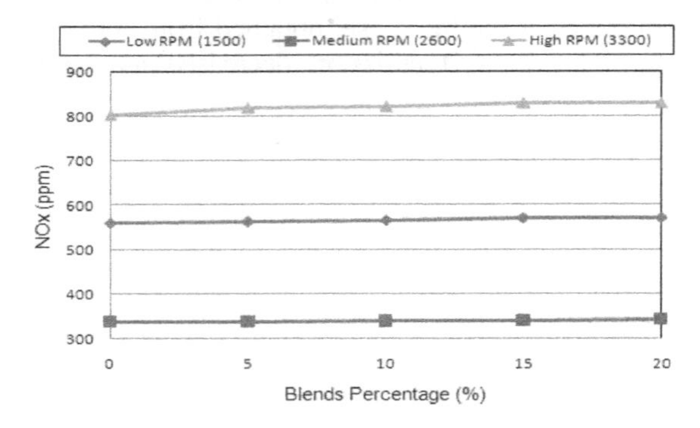

Figure 11: Variation of oxide of nitrogen emissions for different biodiesel blends.

In addition to sunflower bio-diesel various vegetable oils were tested in this study to compare the engine exhaust emissions. Figure 12 shows the effects of standard diesel and bio-diesel fuel on (HC) emission; bio-diesel produces lower HC emission. This may be attributed to the availability of oxygen in bio-diesel, which facilitates better combustion. Also, HC emission of bio-diesel was almost identical. Figure 13 shows the effects of the other bio-diesel on CO_2 emission and comparison with the standard diesel effect.

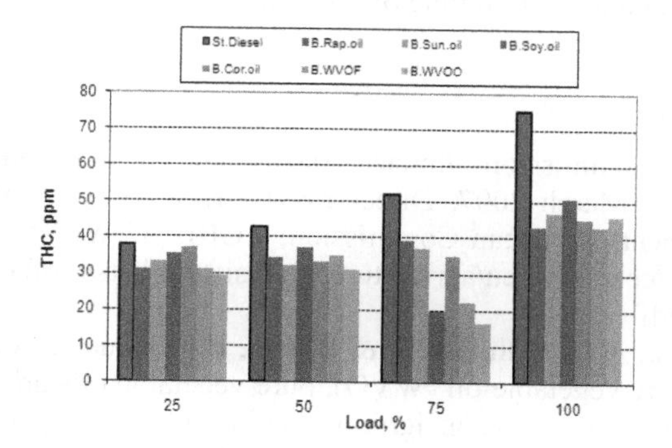

Figure 12: Variation of hydrocarbon emissions with load for fuels tested.

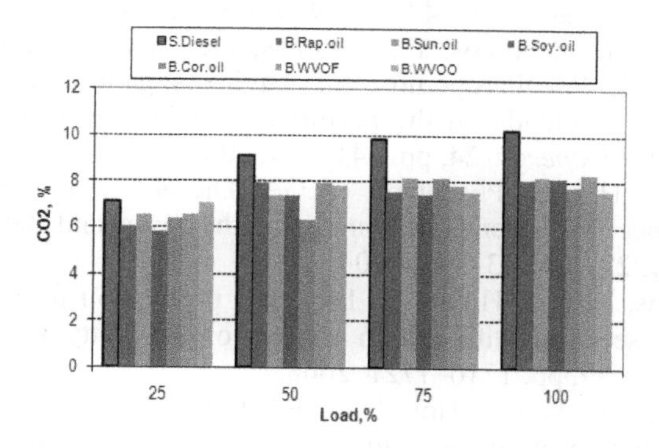

Figure 13: Variation of carbon dioxide emissions with load for fuels tested.

5 Conclusion

The study showed that after testerification of vegetable oils, the kinematic viscosity is reduced from 40 mm^2/s to 5 mm^2/s. For the analyzed samples, the properties were similar in some cases and different in others. Any types of vegetable oils bio-diesel can be used as an alternate and nonconventional fuel to run all types of CI engine. By running biodiesel fuel, the experiential data was showed decreased in almost all the emissions (CO, THC and CO_2) except for NOx. The study has shown that no matter what type of feedstock is used there will be very similar decreases in emissions and performance of the engine. From the combustion analysis it was found the performance of the B20 was as good as that of standard diesel and the difference in the brake power and torque were very small in most cases. RSM proved to be a powerful tool for the optimization of methyl ester production at a fixed temperature. The optimal conditions for the maximum methyl ester yield were found to be at methanol/oil molar ratio of 7.7:1, NaOH catalyst concentration of 1%, reaction temperature 35 °C, rate of mixing 200 rpm and a reaction time of 60 min.

References

[1] European Commission, Impact Assessment of the Renewable Energy Roadmap – March 2007, Directorate-General for Agriculture and Rural Development, European Commission, AGRI, G2/WMD, 2007, Available at: http://ec.europa.eu/agriculture/analysis/markets/biofuel/impact042007/text_en.pdf

[2] Abuhabaya, A., Fieldhouse, J. & Brown, R., Evaluation of properties and use of waste vegetable oil (WVO), pure vegetable oils and standard diesel as used in a compression ignition engine. *Computing and Engineering*, University of Huddersfield, Huddersfield, UK, pp. 71–76, 2010.

[3] Labeckas, G. & Slavinskas, S., The effect of rapeseed oil methyl ester on direct injection diesel engine performance and exhaust emissions. *Energy Conversion Management*, **47**, pp. 1954–1967, 2006.

[4] Kalligeros, S., Zannikos, F., Stournas, S., Lois, E., Anastopoulos, G., Teas, Ch. & Sakellaropoulos, F., An investigation of using biodiesel /marine diesel blends on the performance of a stationary diesel engine. *Biomass & Bioenergy*, **24**, pp. 141–149, 2003.

[5] Raheman, H. & Phadatare, A.G., Diesel engine emissions and performance from blends of Karanja methyl ester and diesel. *Biomass & Bioenergy*, **27**, pp. 393–397, 2004.

[6] Berchmans, H. J. & Hirata, S., Biodiesel production from crude Jatropha curcas L. seed oil with a high content of free fatty acids. *Bioresource Technology*, **99**, pp. 1716–1721, 2008.

[7] Lin, Y.C., Lee, W.J. & Hou, H.C., PAH emissions and energy efficiency of palm-biodiesel blends fueled on diesel generator. *Atmospheric Environment*, **40**, pp. 3930–3940, 2006.

[8] Ahmad, M., Ahmed, S., Ul-Hassan, F., Arashad, M., Khan, M., Zafar, M. & Sultana, S., Base catalyzed transesterification of sunflower oil biodiesel. *African Journal of Biotechnology*, **9**, pp. 8630–8635, 2010.

[9] Yuan, X.Z., Liu, J., Zeng, G.M, Shi, J.G., Tong, J.Y. & Huang, G.H., Optimization of conversion of waste rapeseed oil with high FFA to biodiesel using response surface methodology. *Renewable Energy*, **33**, pp. 1678–1684, 2008.

[10] Gerpen, J.V., Bio-diesel processing and production. *Fuel Process Technology*, **86**, pp. 1097–1107, 2005.

Biodiesel reforming with a $NiAl_2O_4/Al_2O_3$-YSZ catalyst for the production of renewable SOFC fuel

N. Abatzoglou, C. Fauteux-Lefebvre & N. Braidy
Université de Sherbrooke, Canada

Abstract

Biodiesel's contribution as a renewable energy carrier is increasing continuously. Fuel cell market penetration, although slow, is now an irreversible reality. The combination of solid oxide fuel cells (SOFC) with biodiesel offers considerable advantages because it entails both high energy conversion efficiency and near-zero atmospheric carbon emissions. This work is aimed at proving the efficiency of a newly-developed (patent pending), Al_2O_3/YSZ-supported $NiAl_2O_4$ spinel catalyst to steam reform biodiesel. Reforming converts biodiesel into a gaseous mixture, mainly composed of H_2 and CO, used directly as SOFC fuel. The work is performed in a test rig comprising a lab-scale, fixed-bed isothermal reactor and a product-conditioning train. The biodiesel/water mixtures are emulsified prior to their spray injection in the reactor preheating zone, where they are instantaneously vaporized and rapidly brought to the desired reaction temperature to avoid thermal cracking. Reforming takes place at gas hourly space velocities equal to or higher than those in industrial reforming units. The products are analysed by at-line gas chromatography. The results show that biodiesel conversion is complete at steady state. Thermodynamic calculations reveal that the fast reforming reaction reaches chemical equilibrium. The catalyst's performance is very efficient and prevents carbon formation and deactivation.
Keywords: biodiesel, steam reforming, SOFC, nickel, spinel.

1 Introduction

Fuel cell efficiency in converting chemical energy into electricity is significantly higher than that of internal combustion engines. With the world need for sustainable development, via substantial cuts to greenhouse gas emissions and energy costs, the combination of fuel cells with renewable fuels, such as biodiesel, is promising.

Hydrogen (H_2) is the ideal fuel, but solid oxide fuel cells (SOFC) can also be fed by carbon monoxide. Therefore, biodiesel catalytic reforming can serve as a SOFC liquid fuel conversion technology. The main products of biodiesel catalytic reforming are H_2, carbon monoxide (CO) and carbon dioxide (CO_2). Eqn (1) is the core reaction of hydrocarbon steam reforming and (2) is the water gas shift (WGS), a secondary reaction.

$$C_nH_m + nH_2O \rightarrow nCO + (n + m/2)H_2 \quad (\Delta H > 0) \tag{1}$$

$$CO + H_2O \rightarrow CO_2 + H_2 \quad (\Delta H > 0) \tag{2}$$

The purpose of this work is to test a new nickel-alumina spinel (Al_2O_3/YSZ-supported $NiAl_2O_4$) material [1] as catalyst of biodiesel steam reforming.

1.1 Biodiesel reforming

Biodiesel reforming can be represented by the following global reaction (3):

$$C_{18}H_{36}O_2 + 16H_2O \rightarrow 18CO + 34H_2 \quad (\Delta H > 0) \tag{3}$$

Even though biodiesel is well known as a renewable source of fuel for the future, biodiesel steam reforming has not been investigated extensively.

In ref. [2], the authors reported a thermodynamic simulation study of autothermal (ATR) and steam (SR) reforming of various liquid hydrocarbon fuels. They found the highest theoretical conversion efficiency in gasoline, but biodiesel was in the same range (1% lower on average), depicting its feasibility for in-line reforming with fuel cells.

In ref. [3], biodiesel reforming has been simulated and tested in a heat-integrated fuel processor. A commercial precious metal-based catalyst was tested in the fuel processor. These authors obtained 99% conversion in the ATR processor with a steam-to-carbon molar ratio of 2.5, added oxygen, pressure of 2.1 bar, and gas hourly space velocities (GHSV) of 30,000 h^{-1}.

In ref. [4], an experimental study of ATR was performed with platinum (Pt) and rhodium (Rh)-based catalysts synthesized. Hydrogen was produced at temperatures higher than 510°C with a steam-to-carbon molar ratio of 2 and an oxygen-to-carbon molar ratio of 0.4. Coke formed on the catalyst and reactor vessel walls.

Only ATR was investigated in all biodiesel conversion studies reported, both theoretical and experimental. In the experimental studies, only noble metal catalysts were tested. Transition metals (noble and non-noble) are the most catalytically active in hydrocarbon reforming, and noble metals are known to be more resistant but also more expensive [5, 6].

1.2 Liquid hydrocarbon reforming

There are 3 main routes for catalyst deactivation in hydrocarbon reforming: sintering, sulphur poisoning and coking. Sintering is a typical deactivation mechanism for every high temperature catalytic reaction. Sulphur poisoning is expected when fossil fuels are used; this is not the case with biodiesel, which does not contain sulphur moieties. Two main reaction pathways are responsible for coking: the Boudouard reaction (CO disproportionation to C and CO_2), and hydrocarbon cracking. Coke formation mechanisms are different in non-noble and noble metals. Nickel catalysts are prone to coking, because nickel allows carbon diffusion and dissolution which results in whisker carbon formation [7]. Noble metals do not dissolve carbon significantly, but considerable amounts of carbon-rich structures (i.e. graphite layers along the metallic surface) are produced via other carbon deposition mechanisms [7] which lead to coking.

Catalysts used for liquid hydrocarbons reforming reactions are usually deactivated within 100 hours of use [8–10]. In some cases, concentrations closed to theoretical thermodynamic equilibrium can be reached, depending on the catalyst and reaction severity (mainly sufficiently low space velocities).

Noble metal catalysts are deactivated at a slower rate than non-noble metal catalysts. Strohm *et al.* [10] investigated the SR of simulated jet fuel without sulphur and reported constant hydrogen concentrations of 60%vol for 80 hours with a Ceria-Al_2O_3-supported Rh catalyst. The reactions occurred at temperatures below 520°C and water-to-carbon molar ratio of 3. With sulphur added in the feed (35 ppm), the catalyst was deactivated within 21 hours. Ming *et al.* [11] obtained constant H_2 concentrations of 70% over a 73-hour steady state operation for hexadecane steam reforming with an Al_2O_3-supported bimetallic noble metal catalyst and metal-loading <1.5%. The operating conditions were water-to-carbon molar ratio of 2.7 and an operating temperature of 800°C.

In most cases, deactivation occurs within 8 hours when non-noble metal catalysts (mainly nickel) are employed under most reaction severities, with less H_2 in the products [7, 12, 13]. However, Kim *et al.* [14] reported activity of a catalyst over a 53-hour steady state operation, but under favourable conditions. The H_2 concentrations decreased from 72% to 65% with a magnesia-alumina-supported Ni catalyst (Ni/MgO-Al_2O_3) at a temperature of 900°C, GHSV of 10,000 h^{-1} and a water-to-carbon molar ratio of 3. They reported lower deactivation rates for bi-metallic catalyst using noble metal, with the addition of Rhodium to the catalyst.

1.3 Spinel catalyst

Spinel $NiAl_2O_4$ has been studied with fuel cells for internal methane reforming [15]. The catalyst was prepared by solid surface reaction with stoichiometric quantities of Al_2O_3 and nickel oxide (NiO) to form spinel. It was reduced prior to its use; the so-reduced final fresh catalyst was in the form of $Ni_x/Ni_{1-x}Al_2O_{4-x}$. It was reported to be active and relatively stable at a temperature of 800°C in a 250-hour test.

The formation and the stability of spinel and its capacity to be reduced seem to vary according to the reaction undertaken to form spinel, the stoichiometry and Al_2O_3- type. In ref. [16], spinel was produced by solid state reaction with nanometric gamma phase alumina (γ-Al_2O_3) impregnated with nickel nitrate ($Ni(NO_3)_2 \cdot 6H_2O$), at temperatures ranging between 1,000°C and 1,300°C. The authors observed that the catalyst was totally reduced at temperatures higher than 950°C, with a mixture of CO and CO_2 as reducing atmosphere (and oxygen partial pressure of 1×10^{-15} atm). In ref. [17], the authors noted that spinel formed of NiO and α-Al_2O_3 could be reduced at 650°C, in severe reducing conditions with pure H_2 (oxygen partial pressure of 1.9×10^{-18} to 1.0×10^{-20} atm).

2 Experimental work

2.1 Catalyst preparation

The $NiAl_2O_4$-based catalyst tested in this work was produced by the wet impregnation method. Al_2O_3 (mixture of amorphous and γ-Al_2O_3) and YSZ (Y_2O_3-ZrO_2) (50%–50%) support was prepared by mixing the 2 powders mechanically. Al_2O_3 powder size was 40 µm, and YSZ powder size distribution had an upper limit at 20 µm. The Al_2O_3 and YSZ powders were impregnated with an $Ni(NO_3)_2 \cdot 6H_2O$ aqueous solution (targeting a 5% w/w nickel (Ni) load in the final formulation). Water was evaporated, and the resulting impregnated powder was dried overnight at 105°C. The so-dried mixture was calcined at 900°C for 6 hours to form spinel, by a solid state reaction.

2.2 Catalyst characterization

The composition and morphology were analysed by scanning electron microscopy (SEM). SEM was performed using Hitachi field emission gun and energy dispersive X-ray spectroscopy (EDXS) Oxford detector with an ultra-thin ATW2 window.

2.3 Reforming experimentation

A schematic of the reactor is presented in fig. 1. The reactor inner diameter was 46 mm and the catalytic bed length was 60 mm. The catalyst in powder form was dispersed in quartz wool, which was then compacted in the reactor to form a catalytic bed of quartz fibre containing catalyst particulates. This configuration prevented channelling issues and helped obtain a uniform catalytic bed with a small amount of catalyst.

An emulsion-in-water technique was adopted for biodiesel injection. This method was chosen to enhance hydrocarbon/water mixing. The 2 immiscible reactants were emulsified according to a surfactant-aided protocol. The reactants entered at room temperature and were rapidly heated and vaporized in the pre-heating zone maintained at 550°C. The temperature just before the catalyst bed was between 30°C and 45°C below the reaction temperature, depending on operating parameters. Argon served as inert diluent and internal standard for liquid hydrocarbon steam reforming.

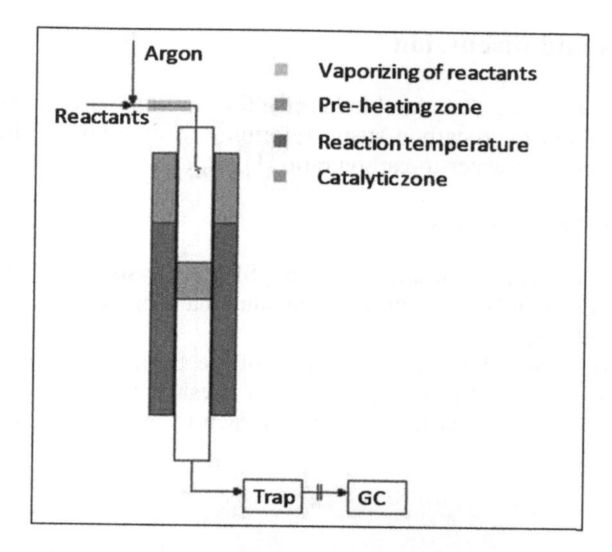

Figure 1: Schematic of the reforming set-up.

The water-to-steam molar ratio was varied between 1.9 and 2.4. Operating temperatures were 700°C and 725°C with GHSV ranging from 5,500 and 13,500 $cm^3_{reac}\ g_{cat}^{-1}\ h^{-1}$ at barometric pressure. Reforming products were analysed by Varian CP-3800 gas chromatography (GC). The exit gaseous flow rate was measured by a flow rate mass meter (Omega FMA-700A). Biodiesel, from used vegetable oil, was produced by a transesterification process developed by Biocarburant PL (Sherbrooke, Qc, Canada; www.biocarburantpl.ca).

Experimental conversion was calculated from:

$$X = \frac{N_{CO out} + N_{CO_2 out} + N_{CH_4 out}}{N_{C_m H_{n_i} n} \times m + N_{Surfactant_{in}} \times Y} \quad (4)$$

with N_i being the total number of moles of component i at the reactor exit or inlet, and Y being the number of carbon atoms in the surfactant. Overall conversion was calculated for liquid hydrocarbon reforming based on the total amount of carbon fed in the reactor. Hydrocarbons were considered to be converted when they were transformed into gaseous products (CO, CO_2 or CH_4). Carbon found in the reactor after the experiment was therefore not considered as converted hydrocarbon.

In the reported tests, the reactor exit concentrations of H_2, CO, CO_2 and CH_4 were compared to theoretical thermodynamic equilibrium concentrations to determine if equilibrium was reached. Thermodynamic equilibrium concentrations were calculated with FactSage software on the basis of Gibbs energy minimization.

3 Results and discussion

The catalyst presented here for biodiesel reforming has already proved to be efficient for liquid hydrocarbon steam reforming at high GHSV and relatively low temperatures and water-to-carbon ratio [1].

3.1 Catalyst characterization

The catalyst formulation was analyzed using SEM analysis. The targeted catalyst form is $NiAl_2O_4$ spinel on the surface of an alumina support without any metallic nickel or nickel oxide.

Surface SEM and SEM-EDXS analyses of the fresh catalyst are reported in ref. [1]. Figure 2 is a SEM analysis of the fresh catalyst surface and more particularly of the Al_2O_3 surface which is known to be the main support of the spinel phase.

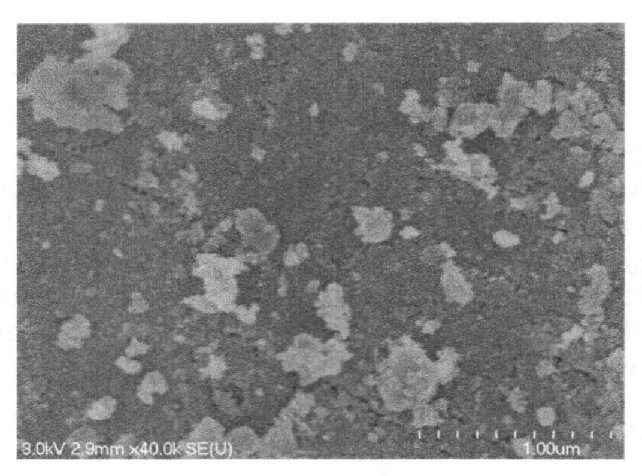

Figure 2: SEM micrograph of the Al_2O_3 surface.

3.2 Steam reforming results

3.2.1 Measurement errors

The errors associated with concentration data obtained by GC appear in table 1. They were calculated with an external standard.

Table 1: Gas concentration measurement errors.

Gas	Standard gaseous concentration (%)	Absolute error (on % concentration of the standard)	Relative error (%)
H_2	55.16	0.46	0.83
CO	19.70	0.21	1.05
CO_2	6.96	0.38	5.45
CH_4	2.08	0.04	1.87
Ar	16.10	0.22	1.37

In addition to GC concentration measurement errors, the mass flow meter for quantifying exit gas flow introduced a second error in the conversion calculations. The accuracy of the mass flow meter was 1%. Maximum and minimum values were therefore calculated for each conversion, with extreme values for concentrations and flow rates based on known error and accuracy.

3.2.2 Biodiesel steam reforming

Table 2 lists the conditions of 3 different biodiesel reforming test runs with the associated overall conversion calculated.

Table 2: Biodiesel reforming test run description.

Run	1	2	3
Temperature (°C)	700	725	725
Catalyst weight (g)	5.0	3.0	3.0
Run time (h)	3	4	2
GHSV ($cm^3g^{-1}h^{-1}$)	8,700	5,500	13,500
H2O/C[a] (mol/mol)	1.9	1.9	2.4
Conversion (± 3%)	88	100	85

[a]Water-to-carbon (H_2O/C) ratio calculated including surfactant.

Dry gaseous concentrations at the reactor exit are presented in fig. 3. Concentrations were stable for the entire reaction time with no catalyst deactivation observed.

Temperature increase and flow rate decrease would obviously lead to 100% conversion. It can also be observed that an increase of GHSV decreases conversion, even at a higher H_2O/C ratio. This reduction of conversion is associated with reaction kinetics. Fig. 3 compares the theoretical equilibrium and experimental concentrations of the dry gas at the reactor exit.

These preliminary data are indicative of the ability of this catalytic formulation to efficiently steam reform commercial biodiesel. The catalyst is not poisoned by sulphur (not present in biodiesel in detectable quantities), and since carbon formation is insignificant, the only remaining catalyst deactivation mechanism is sintering. Although the extent of the performed tests is not sufficient to allow us to evaluate such a mechanism, $NiAl_2O_4$ thermal mobility is much lower (insignificant at reaction conditions) than that of metallic Ni. Thus, the expected life cycle of the proposed catalyst is considerably longer than any other metallic Ni-based formulation.

High GHSV, which give complete biodiesel conversion, are indicative of a rather surface reaction kinetics-controlled process. However, additional experiments are needed, in conditions under which the reaction does not reach chemical equilibrium, in order to evaluate the kinetic parameters (mainly activation energy) as well as the mass transfer and chemical reaction resistances.

The concentrations for run 2 were equal to those at chemical thermodynamic equilibrium. In run 1, even if conversion was not complete, the concentrations were near equilibrium. It should be noted that for biodiesel reforming below

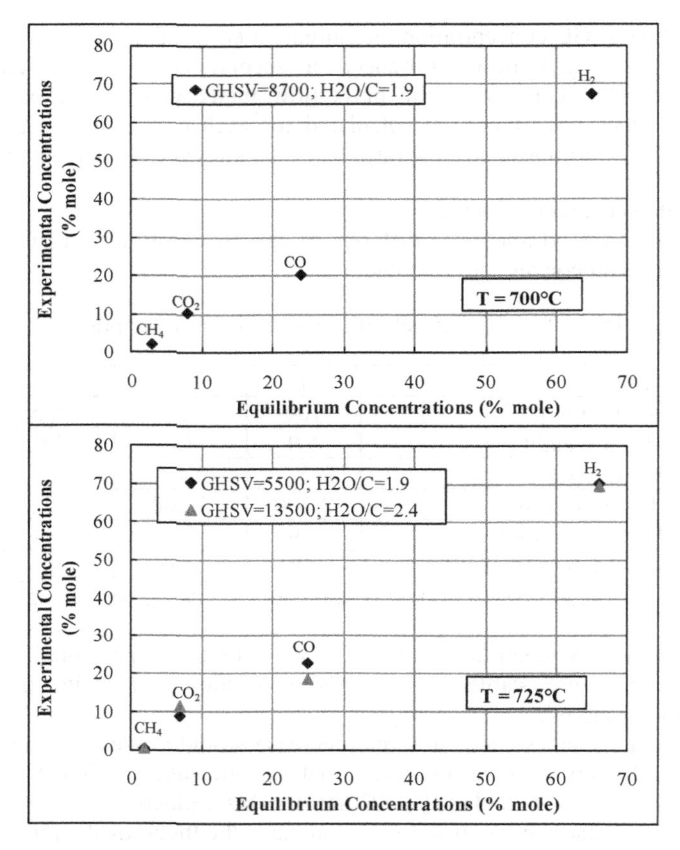

Figure 3: Experimental vs. theoretical concentrations in biodiesel reforming product (errors in values are less than 1% in all cases).

700°C, theoretical equilibrium concentrations predict the presence of significant amounts of methane and coke formation if the water-to-carbon ratio in reactants is not higher than stoichiometric ratio.

3.3 Used catalyst characterization

Figure 4 is a SEM micrograph of an Al_2O_3 particulate of the $NiAl_2O_4$ catalyst used in run 2 of the biodiesel reforming test and comparison with fig. 2, which is the same for the fresh catalyst, proves that there was no significant carbon deposition on the surface. Some carbon whiskers were found on an extent lower than 5% of the surface; this is, however, expected because of local surface nanoheterogeneities and the possibility that some NiO on the surface was not transformed into $NiAl_2O_4$ which could form Ni during SR reactions.

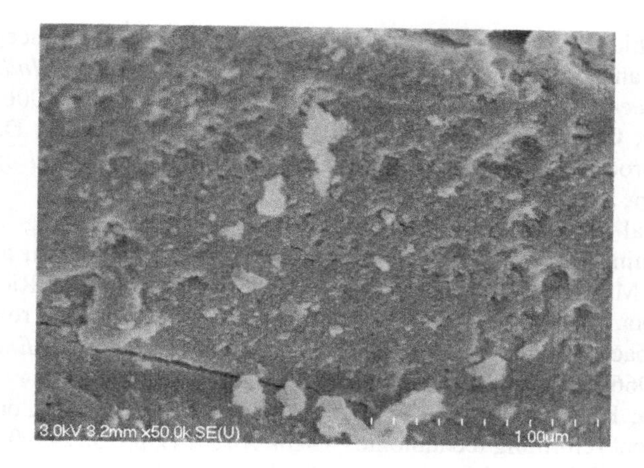

Figure 4: SEM picture of the catalyst after run 2.

4 Conclusion

An Al_2O_3/YSZ-supported $NiAl_2O_4$ catalyst has been tested efficiently in biodiesel SR. 100% conversion was obtained at relatively low severity conditions. Increasing GHSV above 10,000 $cm^3g^{-1}h^{-1}$ decreased conversion, but dry concentrations of the exit gas were still near equilibrium. No catalyst deactivation was encountered. There was no observable carbon on the surface of the catalyst used in these conditions, even with a water-to-carbon ratio lower than 2.

Acknowledgements

The authors are indebted to SOFC Network Canada and the Agricultural Biomass Innovation Network (ABIN) for funding related to this project. The financial contribution of the National Science & Engineering Research Council (NSERC) of Canada through Discovery Funding and Students Awards is also acknowledged along with the Le Fonds québécois de la recherche sur la nature et les technologies (FQRNT) for Students Awards. Biodiesel was kindly provided by Biocarburant PL. Many thanks are due to Carmina Reyes Plascencia and Henri Gauvin for their technical support and to Sonia Blais and Stéphane Gutierrez for their help in catalyst characterization. Finally, special thanks to Ovid Da Silva for reviewing the manuscript.

References

[1] Fauteux-Lefebvre, C., Abatzoglou, N., Blanchard, J. & Gitzhofer, F., Steam reforming of liquid hydrocarbons over a nickel-alumina spinel catalyst. *Journal of Power Sources*, **195(10)**, pp. 3275–3283, 2010.

[2] Specchia, S., Cutillo, A., Saracco, G. & Specchia, V., Concept study on ATR and SR fuel processors for liquid hydrocarbons. *Industrial and Engineering Chemistry Research*, **45(15)**, pp. 5298–5307, 2006.
[3] Kraaij, G.J., Specchia, S., Bollito, G., Mutri, L. & Wails, D., Biodiesel fuel processor for APU applications. *International Journal of Hydrogen Energy*, **34(10)**, pp. 4495–4499, 2009.
[4] Ospinal-Jiménez, M., Hydrogen production study using autothermal reforming of biodiesel and other hydrocarbons for fuel cell applications, 2006, Master of Science degree thesis, University of Puerto Rico.
[5] Ibarreta, A.F. & Sung, C., Optimization of Jet-A fuel reforming for aerospace applications. *International Journal of Hydrogen Energy*, **31(8)**, pp. 1066–1078, 2006.
[6] Lucka, K. & Kohne, H., Challenges in diesel reforming: Comparison of different reforming technologies. *Clean Air*, **7(4)**, pp. 381–390, 2006.
[7] Alvarez-Galvan, M.C., Navarro, R.M., Rosa, F., Briceno, Y., Gordillo Alvarez, F. & Fierro, J.L.G., Performance of La,Ce-modified alumina-supported Pt and Ni catalysts for the oxidative reforming of diesel hydrocarbons. *International Journal of Hydrogen Energy*, **33(2)**, pp. 652–663, 2008.
[8] Cheekatamarla, P.K. & Lane, A.M., Catalytic autothermal reforming of diesel fuel for hydrogen generation in fuel cells: I. Activity tests and sulfur poisoning. *Journal of Power Sources*, **152(1-2)**, pp. 256–263, 2005.
[9] Rosa, F., Lopez, E., Briceno, Y., Sopena, D., Navarro, R.M., Alvarez-Galvan, M.C., Fierro, J.L.G. & Bordons, C., Design of a diesel reformer coupled to a PEMFC. *Catalysis Today*, **116(3)**, pp. 324–333, 2006.
[10] Strohm, J.J., Zheng, J. & Song, C., Low-temperature steam reforming of jet fuel in the absence and presence of sulfur over Rh and Rh-Ni catalysts for fuel cells. *Journal of Catalysis*, **238(2)**, pp. 309–320, 2006.
[11] Ming, Q., Healey, T., Allen, L. & Irving, P., Steam reforming of hydrocarbon fuels. *Catalysis Today*, **77(1-2)**, pp. 51–64, 2002.
[12] Gardner, T.H., Shekhawat, D., Berry, D.A., Smith, M.W., Salazar, M. & Kugler, E.L., Effect of nickel hexaaluminate mirror cation on structure-sensitive reactions during n-tetradecane partial oxidation. *Applied Catalysis A: General*, **323**, pp. 1–8, 2007.
[13] Gould, B.D., Tadd, A.R. & Schwank, J.W., Nickel-catalyzed autothermal reforming of jet fuel surrogates: n-Dodecane, tetralin, and their mixture. *Journal of Power Sources*, **164(1)**, pp. 344–50, 2007.
[14] Kim, D.H., Kang, J.S., Lee, Y.J., Park, N.K., Kim, Y.C., Hong, S.I. & Moon, D.J., Steam reforming of n-hexadecane over noble metal-modified Ni-based catalysts. *Catalysis Today*, **136(3-4)**, pp. 228–234, 2008.
[15] Kou, L. & Selman, J., Activity of NiAl2O4 catalyst for steam reforming of methane under internal reforming fuel cell conditions. *Electrochemical Society Proceedings*, **99(19)**, pp. 640–646, 1999.
[16] Huang, Z.R., Jiang, D.L., Michel, D., Mazerolles, L., Ferrand, A., di Costanzo, T. & Vignes, J.L., Nickel-alumina nanocomposite powders

prepared by novel in situ chemical reduction. *Journal of Materials Research*, **17(12)**, pp. 3177–3181, 2002.

[17] Jiong, Y.S., Kou, L., Nash, P. & Selman, J.R., Behavior of nickel aluminate spinel under reducing conditions. *Electrochem. Soc. Proc.*, pp. 456–68, 1997.

On the future relevance of biofuels for transport in EU-15 countries

A. Ajanovic & R. Haas
Energy Economics Group, Vienna University of Technology, Austria

Abstract

The discussion on the promotion of biofuels in the European Union (EU) countries is ambiguous: benefits like reduction of greenhouse gas emissions and increase of energy supply security are confronted with high costs and bad ecological performance. On the one hand, the EU has set the goal of reaching 10% biofuels by 2020. On the other hand, there are continuous persisting discussions to undermine this goal. The core objective of this paper is to investigate the market prospects of biofuels for transport in the EU-15 in a dynamic framework till 2030. While the economic prospects for the first generation of biofuels are rather promising – cost-effectiveness under current tax policies exists already – their potential is very restricted especially due to limited crops areas. Moreover, the environmental performance of first generation biofuels is currently rather modest. Second generation biofuels will – in a favourable case – enter the market between 2020 and 2030. However, their full potential will be achieved only after 2030.
Keywords: biofuels, costs, potential, CO_2-emissions.

1 Introduction

The discussion on the promotion of biofuels is ambiguous: benefits like the reduction of greenhouse gas emissions and increase of energy supply security are confronted with high costs and bad ecological performance. Great hopes are currently put on biofuels second generation. The major advantage of the second generation of biofuels is that they can also be produced from resources such as ligno-cellulose based wood residues, waste wood or short-rotation copies, which are not dependent on food production-sensitive crop areas.

The core objective of this paper is to investigate the market prospects of biofuels for transport in the EU-15 in a dynamic framework till 2030. This work extends the analysis conducted by Ajanovic and Haas [1]. With respect to the literature the most important analyses are summarized by Panoutsou *et al.* [2] and Ajanovic and Haas [3].

The following categories are considered:

- Biofuels 1st generation: biodiesel from rapeseed, sunflowers, soybeans (BD-1); bioethanol from maize, wheat, sugar beet (BE-1); biogas (BG-1) from manure, grass and green maize.
- Biofuels 2nd generation: biodiesel from biomass-to liquids (BTL) with Fischer-Tropsch process (BD-2); bioethanol from lingo-cellulose (BE-2); biogas (BG-2) from synthetic gas from biomass.

These biofuels are analysed with regard to potentials, costs and market prospects, and the environmental impacts. This analysis is based on:

- possible developments of fossil energy price levels and energy demand;
- technological learning effects (based on global developments);
- environment, energy and transport policies on EU level.

2 Method of approach

The method of approach of our analysis consists of the following major steps:

2.1 Assumptions

Major assumptions for the modelling analysis are:

- Increases in fossil fuel prices are based on IEA [4].
- The development of alternative fuel costs is based on international learning rates of 25% and national learning rates of 15% regarding the investment costs of these technologies.
- International learning corresponds to world-wide quantity developments in the Reference Scenario (RS) and the Alternative Policy Scenario (AS) in IEA [4] up to 2030.
- All cost figures are in prices of 2008.
- No explicit carbon costs are included.
- Regarding the future land use, it has been assumed that maximal 30% of arable land, 10% of pasture land, 10% of meadows and 3% of wood and forest wood residues could be used for feedstock production for biofuels by 2030. An additional 5% of wood industry residues could be used for biofuels production.

2.2 Calculation of biofuel costs

Next the biofuel production costs are calculated. The following components are considered to calculate the costs of biofuels (see also Ajanovic and Haas [1]):

- Net feedstock costs (C_{FS}).
- Gross conversion costs (GCC).
- Distribution and retail costs (DC).
- Subsidies for biofuels (Sub_{BF}).

Firstly, the feedstock costs are identified for every year as the minimum production costs of all possible feedstocks considered for a specific area category (e.g. crop area) as:

$$C_{FS_t} = \mathrm{Min}(C_{FS_{i\_t}}; i = 1, ..., n)$$

where n is the number of possible feedstocks.

Finally, the total biofuel production costs (C_{BF}) for year t are calculated as follows (note that in these analyses no explicit carbon costs are included):

$$C_{BF} = C_{FS} + GCC + DC - Sub_{BF}$$

2.3 Considering technological learning

Future biofuel production costs will be reduced through technological learning. Technological learning is illustrated for many technologies by so-called experience or learning curves. In the present model, specific investment costs $IC_t(x)$ are split up into a part that reflects the costs of conventional mature technology components $IC_{Con\_t}(x)$ and a part for the new technology components $IC_{New\_t}(x)$:

$$IC_t(x) = IC_{Con\_t}(x) + IC_{New\_t}(x)$$

For $IC_{Con\_t}(x)$, no more learning is expected. For $IC_{New\_t}(x)$, the following formula is used to express an experience curve by an exponential regression:

$$IC_{New\_t}(x) = a \cdot x_t^{-b}$$

where:

$IC_{New\_t}(x)$	Specific investment cost of new technology components (€/kW)
x_t	Cumulative capacity up to year t (kW)
b	Learning index
a	Specific investment cost of the first unit (€/kW)
$IC_{Con\_t}(x)$	Specific investment cost of conventional mature technology components (€/kW).

2.4 Maximum additionally usable areas

Then, for every area category considered, the maximum additional feedstock area per year (A_{FS_ADDt}) is calculated as:

$$A_{FS\_ADD\_t} = \phi(A_{FS\_MAX\_t} - A_{FS\_t-1})$$

with ϕ the maximum percentage to be added or reduced per year.

2.5 Actual additional areas used

Additional feedstock areas are used for biofuels under the following conditions:

$$A_{FS\_t} = A_{FS\_t-1} + A_{FS\_Addt} \left| C_{BFt}(C_{FSt})[1 + \tau_{BF}] < p_{FFt}[1 + \tau_{FF}] \right.$$

where

τ_{BF} tax on biofuels

τ_{FF} tax on fossil fuels

p_{FF} price of fossil fuels (excl. tax).

In contrast, the feedstock area is reduced when

$$A_{FS\_t} = A_{FS\_t-1}(1-\phi)\left|C_{BFt}(C_{FSt})[1+\tau_{BF}] > p_{FFt}[1+\tau_{FF}]\right.$$

2.6 Assigning feedstock areas to biofuel categories

Feedstocks as well as feedstock areas may also be used for different biofuel categories. For example, some crop areas are suitable for oilseeds for 1st generation biodiesel (BD-1), for wheat for 1st generation bioethanol (BE-1) and for corn stover for 2nd generation bioethanol (BE-2). In this case the feedstock potentials or the feedstocks' area are dedicated to the biofuels category which leads to the cheapest production costs per kWh biofuel:

$$C_{FS_t} = \text{Min}(C_{FS_{j\_t}}; j = 1,\ldots,m)$$

where m is the number of possible biofuels categories.

3 Potential

There follows a presentation of the results of cost development and corresponding quantities produced for 1st and 2nd generation biofuels in EU-15 up to 2030. These alternative energy carriers are based on bioenergy resources. An increasing use of biomass in the future in Europe could raise basically two questions: (i) the use of biomass requires large amounts of land which otherwise could be used for other purposes (e.g. food production); (ii) increasing biomass production might be in contradiction with sustainable issues.

The total land area in EU-15 is about 313 million hectares. This total land area could be divided in five groups: arable land (23%), permanent crop (3%), permanent meadows and pastures (16%), forest area (39%) and other land (19%) (see fig. 1).

However, with the second generation of biofuels, other land areas such as meadows, pastures and forest area could also be used for biofuels production, so that total land potential for alternative energy carriers could be significantly higher.

Due to the EU targets regarding biofuels share in total transport fuel consumption, it could be expected that total energy from biofuels by 2030 could be significantly higher than now.

As shown in fig. 1, the total energy from biomass in 2030 could be more than three times higher than now, 720 TWh. After about 2023 a significant and continuously increasing share of the 2nd generation bioethanol could be noticed. The share of 2nd generation biodiesel could be significant starting from 2027 due to the lower costs than conventional diesel (see fig. 2).

The increasing biofuels production based on domestic produced feedstock will occupy additionally land use (see fig. 3). However, for 2nd generation biofuels mainly non-crop area dependent resources will be used. These are: straw, waste wood and wood residues from the industry.

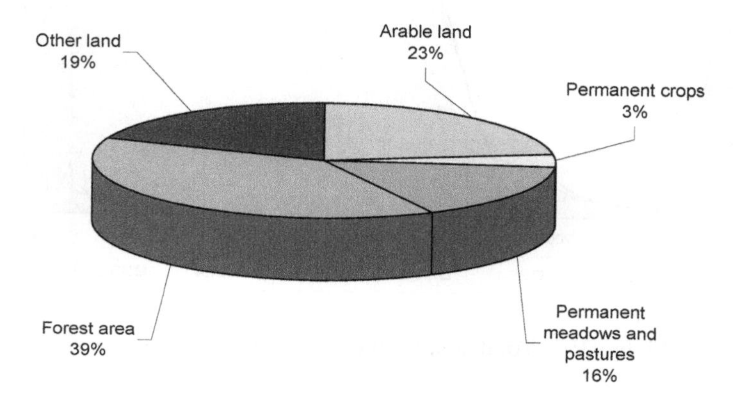

Figure 1: Land area in EU-15, 2008 [5].

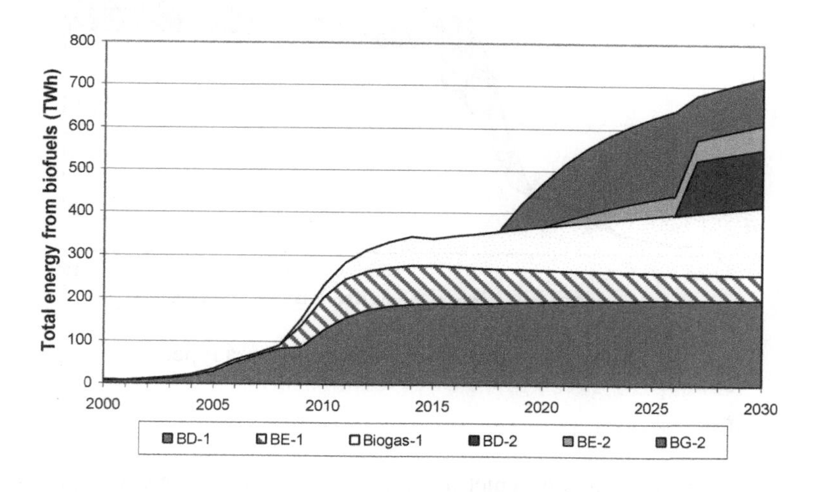

Figure 2: Total energy from biofuels by biofuels category.

Due to the switch to the second generation biofuels, up to 2030 also significant poplar areas will be used for feedstock production (see fig. 4). The total land area for biofuels production by 2030 will be 64.2 million hectares.

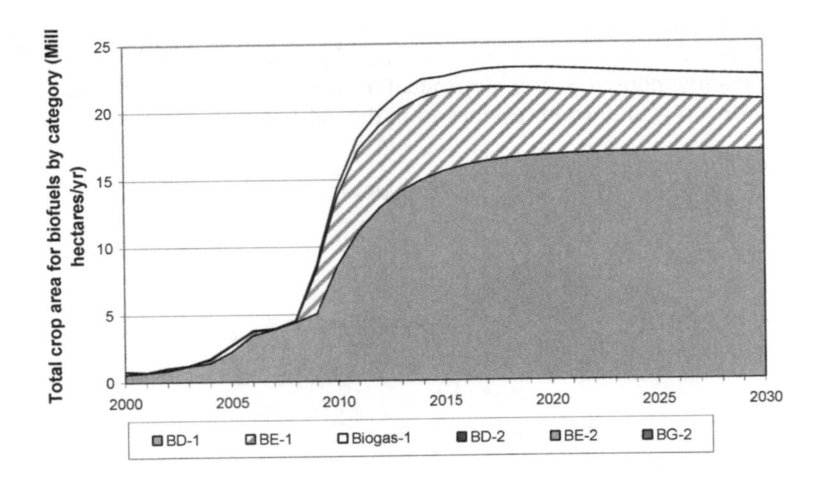

Figure 3: Total area for biofuels by biofuels category.

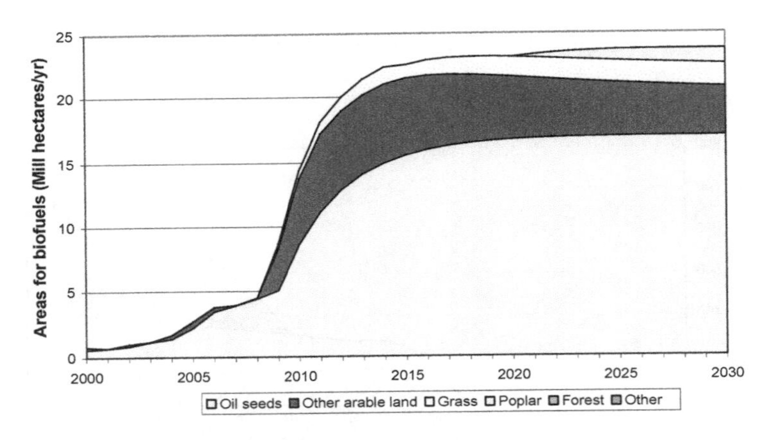

Figure 4: Areas for biofuels by area type.

4 Costs

The following figs 5 and 6 depict the corresponding development of production costs (inclusive and exclusive of 20% VAT) and the prices of fossil fuels, gasoline and diesel, inclusive and exclusive of taxes.

As can be seen from fig. 5 and fig. 6, the costs of 1st generation biofuels are decreasing only slightly even in the most favourable scenario. The major cost reduction is caused by learning effects for capital costs.

As described above, these learning effects are trigged mainly by international learning. They are in the present work based on the quantities development in the Referent (RS) and Alternative Policy Scenario (AS) of IEA [4].

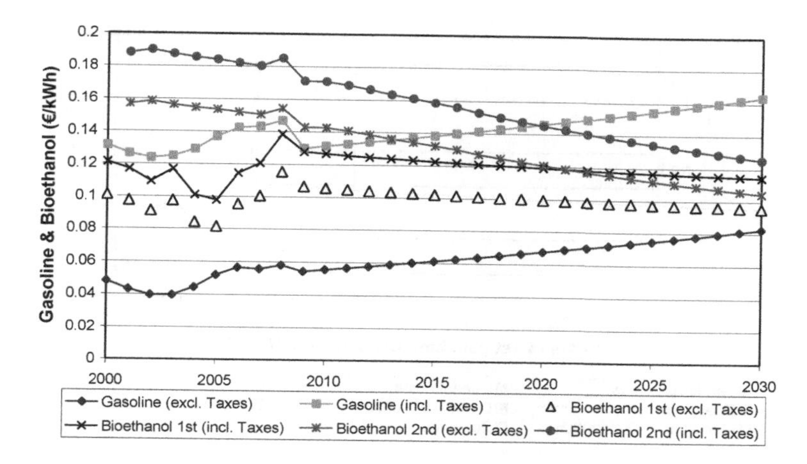

Figure 5: Price versus costs of gasoline and bioethanol.

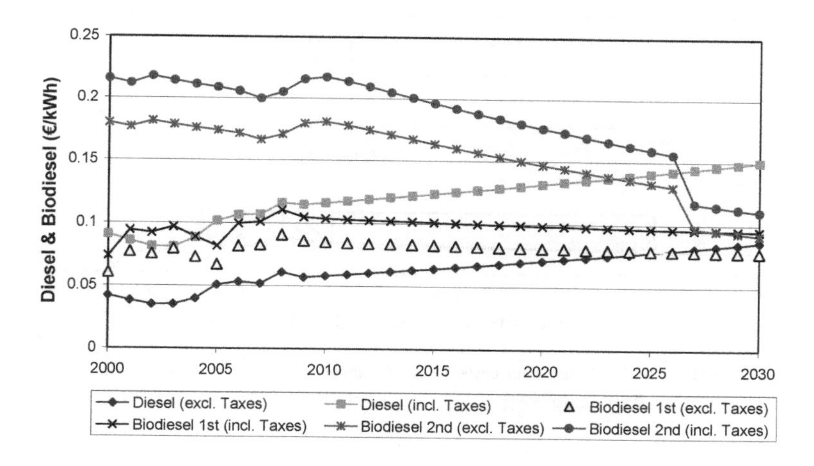

Figure 6: Price versus costs of diesel and biodiesel.

The major results of this analysis are: (i) 2nd generation bioethanol will become competitive when including current tax schemes by about 2020 (see fig. 5); (ii) Biodiesel (BD-2) will compete with fossil diesel only close after 2025 (see fig. 6); (iii) Yet, if no taxes are considered, neither 1st nor 2nd generation bioethanol will be cheaper than fossil fuels before 2030. Close before 2030 biodiesel 1st generation could become competitive with fossil diesel without tax exemptions.

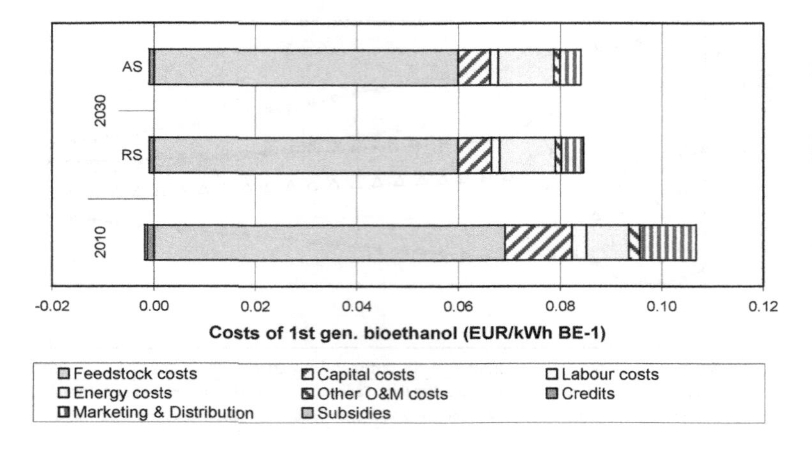

Figure 7: Costs of 1st generation bioethanol (2010 vs. 2030).

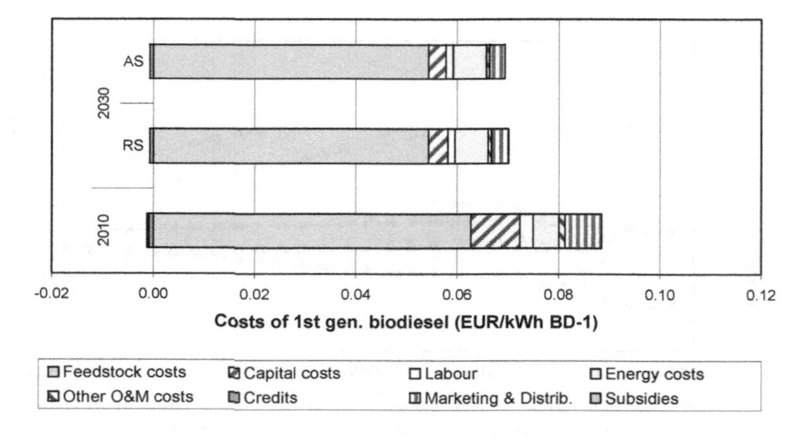

Figure 8: Costs of 1st generation biodiesel (2010 vs. 2030).

Figures 7 and 8 depict the underlying detailed cost structures. It can be seen that the largest part of the total biofuels costs are feedstock costs. In the future, the major cost reduction could be caused by capital costs. But the actual cost differences between RS and AS are rather small.

5 Environmental performance

A very sensitive issue with respect to the future relevance of biofuels is their energetic and environmental performance.

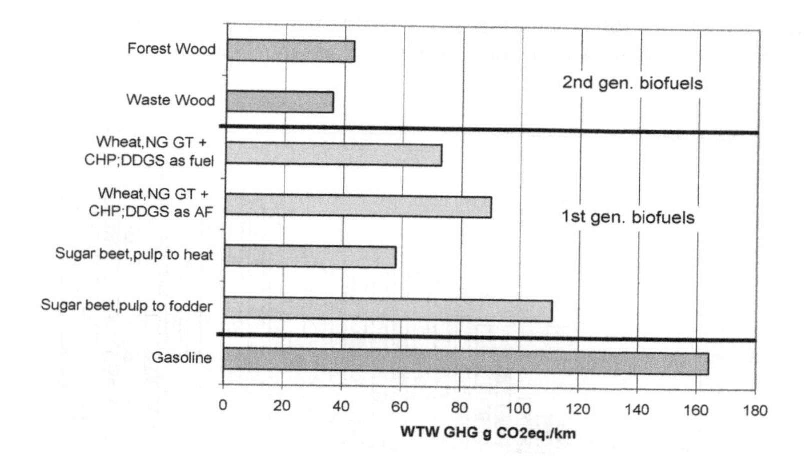

Figure 9: Bioethanol: total WTW GHG emissions [6].

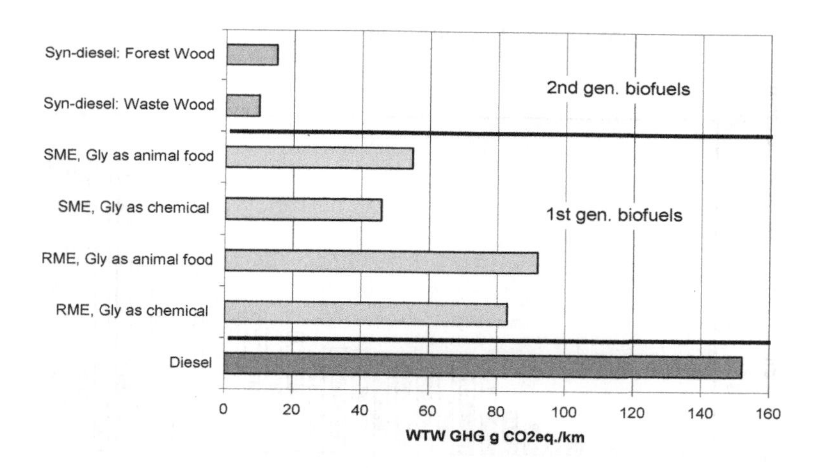

Figure 10: Biodiesel: total WTW GHG emissions [6].

The range of the GHG emissions is very wide due to the different production technology, different feedstocks and the way of using by-products. As shown in fig. 9 and fig. 10, conventional biofuels have moderate reduction of GHG emissions. Higher GHG emission reductions could be achieved in the case of by-products being used as fuel instead of as animal feed. However, GHG emission reductions for the 2nd generation biofuels could be much higher, mostly because these processes use part of the biomass intake as fuel and therefore involve less input of fossil energy [7].

The CO_2 emissions profile of biofuels production depends very much on the type of feedstock used and the production process. With the increasing use of biofuels in EU-15 total emission from biofuels will be in 2030 significantly higher than now, about 83 million tons $CO_{2\text{-eq}}$ (see fig. 11).

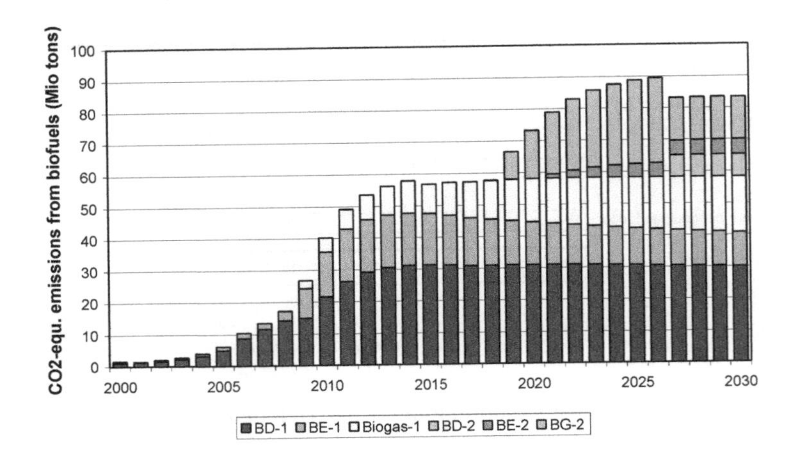

Figure 11: $CO_{2\text{-eq}}$ emissions from biofuels [7].

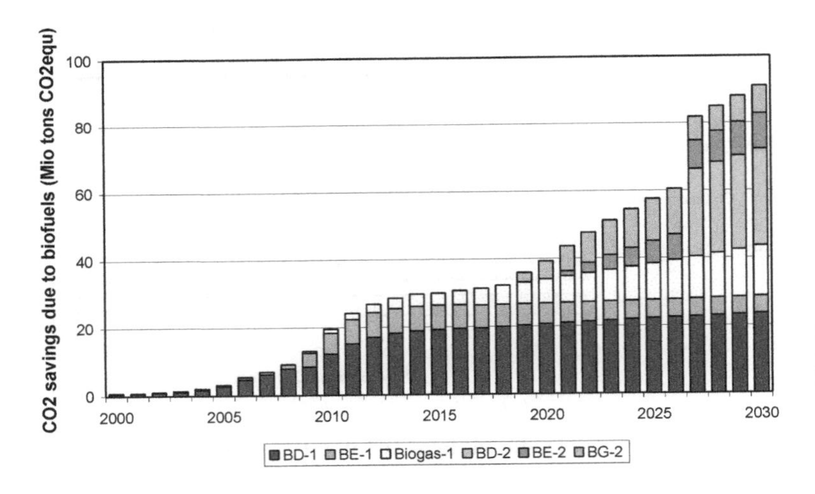

Figure 12: CO_2 savings in EU-15 due to biofuels [7].

However, using biofuels, considerable CO_2 saving in EU-15 could be noticed (see fig. 12). An increase in CO_2 saving after 2026 is due to the increasing share of biofuels 2nd generation.

In this context, it is very important to state that it has to be ensured by monitoring and certification processes that the ecological performance of biofuels 1st generation improves continuously.

6 Conclusions

The major conclusions are:

- Under current policy conditions – mainly exemption of excise taxes – the economic prospects of biofuels 1st generation in Europe are rather promising; the major problems of biofuels 1st generation are lack of available land for growing proper feedstocks and the modest ecological performance; the indigenous potential for BF-1 in EU-15 are limited at a level of about 2 to 3 times the volume of today (without endangering food supply and without imports of feedstocks for biofuels like palm oil).
- The environmental performance of the 1st generation biofuels is currently rather modest.
- Biofuels 1st generation will reach their maximal potential by about 2020. Since up to 2030 they are still cheaper than 2nd generation biofuels, they will remain in the market at least until 2030 without significant reductions.
- Large expectations are put into advanced 2nd generation biofuels production from ligno-cellulosic materials like whole plants, wood and wood residues; so the major advantage of BF-2 is that the potential will be significantly higher at levels of more than ten times of today's BF production; (vi) Regarding the future costs of BF-2 it can be stated that in a favourable case by 2030 they will be close to the costs of BF-1; so by 2030 in Europe neither for BF-1 nor for BF-2 significantly lower costs can be expected. Yet, if prices of fossil fuels continue to increase at least slightly given current tax policies, BF-1 will become competitive already in the coming years, BF-2 about a decade later.
- 2nd generation biofuels will – in a favourable case – enter the market between 2020 and 2030. However, their full potentials will be achieved only after 2030.

References

[1] Ajanovic, A. & Haas, R., Economic challenges for the future relevance of biofuels in transport in EU-countries. *Energy*, **35**, pp. 3340–3348, 2010.
[2] Panoutsou, C., Eleftheriadis, J. & Nikolaou, A., Biomass supply in EU27 from 2010 to 2030. *Energy Policy*, **37**, pp. 5675–5686, 2009.
[3] Ajanovic, A. & Haas, R., The future relevance of alternative energy carriers in Austria. *IEESE-5*, Denizli, Turkey, 2010.
[4] IEA. World Energy Outlook 2009. International Energy Agency, Paris, 2009.
[5] FAOSTAT, http://faostat.fao.org, 2010.
[6] CONCAWE, EUCAR, JRC EU Commission. Well-to-Wheels report, 2007.
[7] Ajanovic, A., ed., Deriving effective least-cost policy strategies for alternative automotive concepts and alternative fuels-ALTER-MOTIVE, www.alter-motive.org, 2011.

Sustainability of combustion and incineration of renewable fuels: the example of Sweden

R. Sjöblom[1,2] & S. Lagerkvist[3,4]
[1]*Luleå University of Technology, Sweden*
[2]*Tekedo AB, Sweden*
[3]*Umeå University, Sweden*
[4]*Land and Environment Court at the Umeå District Court, Sweden*

Abstract

According to the statistics at the European Commission, Sweden is the champion by far in Europe in terms of renewable energy. It comprised around 45% of the total in the year 2008. This position has been reached by a combination of natural resources, political determination and technology development. A major contributor to this is the extensive utilization of district heating which amounts to around 50 TWh per year, and which covers about half of the total need for industrial and domestic buildings. The district heating is based mainly on combustion of bio fuels together with waste and some peat. This practice is generally very positive from a sustainability perspective for the following reasons: (1) bio fuels are renewable, and so is peat, although over a longer time span, (2) waste is being recovered for energy purposes, and (3) ash material is, in many cases, re-circulated and recycled. However, sustainability is not only about total percentages, but also depends on the quality in the processes, especially in terms of qualification of fuels and ashes and the associated possibilities for more efficient combustion and incineration processes as well as ash utilization. Efficiency in this regard of course also includes protection of health and the environment. These aspects are explored in a technical as well as a legal perspective, and some possibilities for further development and improvement are identified and discussed. The compilation and analyses are based on more than ten years of research reports (mostly in Swedish) financed by branch research organizations and similar, see Acknowledgements.
Keywords: sustainable, combustion, incineration, bio fuels, waste, ash, Sweden.

1 Introduction, purpose and objective

1.1 Introduction

It might be tempting to assume that energy from renewable sources in general, and not least thermo-chemical energy, is sustainable by definition. Nonetheless, energy beneficiation, however renewable, would still not be sustainable, at least not entirely, if it were associated with – especially permanent – detriment to health and the environment and/or the consumption of limited resources other than energy resources.

Moreover, since renewable energy can usually replace non-renewable energy resources, it is also essential with regard to sustainability that the utilization is efficient.

Flow of matter and energy, as in the case of combustion and incineration, might be referred to as metabolism, and such a system might be considered as part of the anthroposphere [1]. Thus, when sustainability is to be assessed, it is the complete system from preparation of the fuel to and including the beneficiation or disposal of the ash that needs to be analysed.

The performance of such a process system is quite complex and has been the subject of intense research and development for the last couple of decades by many researchers and engineers and through financing by many entities, including branch research organizations (see Acknowledgements).

Moreover, the process system has undergone far-reaching changes over the same period of time. As will be dealt with further in section 4, most of the waste is nowadays sorted and recycled, whilst most of it was just land-filled as received twenty years ago.

External prerequisites for the process system have also changed substantially. Residues may nowadays in certain cases be regarded as by-products, and waste might be recycled to become products. For products, a novel EC regulation applies – REACH.

1.2 Purpose and objective

Obviously, some complex but important relations exist in the process system just described, several of which are closely related to sustainability.

The purpose of the present work is to describe such features and to discuss different alternatives and their implications. The analyses and the conclusions are to be based on brief compilations of the following:

- a brief background on thermo-chemical energy from renewable sources in Sweden in a European perspective,
- some legal aspects including the differentiation between waste and products, and
- some relations between fuel quality, furnace, flue gas cleaning and ash.

Classification of ash under the EC directive of waste has been dealt with in ref. [2] and such evaluations also form the basis for assessment of the suitability of ash for various purposes and destinations. Work is in progress on utilization of ash for remediation of mine tailings rich in sulfides as well as on sorting of waste before combustion and incineration, and those results will be published later. The analyses in the present paper focus on qualification of fuel as waste or as a product together with the associated prerequisites for combustion and incineration.

2 Thermo-chemical renewable energy utilization in Sweden

According to the Internet page of the Government Agency *Statistics Sweden*, Sweden had a population of 9,555,893 inhabitants at the end of the year 2012, corresponding to a population density of 18 inhabitants per square kilometre.

The climate is cold and the industry is rather energy intensive, thus the energy generation and consumption is high (see fig. 1).

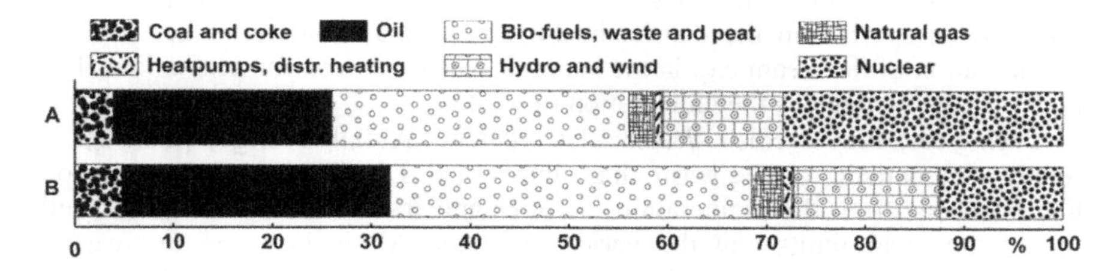

Figure 1: Energy input in Sweden for the year 2011. Alternative A refers to nuclear energy figured as total heat and corresponds to 62 MWh per capita. Alternative B is based on nuclear energy figured as electricity output, and this corresponds to 50 MWh per capita.

As can be seen in the figure, Sweden is high in nuclear electric energy generation, and actually champion in the world with 6.4 MWh per capita. The sustainability of nuclear energy with regard to decommissioning and waste management is dealt with in ref. [4].

However, the present paper focuses on the biofuels, waste and peat portions in fig. 1. According to ref. [5], annex 1, the share of energy from renewable sources in gross final consumption was 39.8% in the year 2005, more than 7% more than Latvia and Finland which were second and third, respectively within the EC. The share has increased since, and the goal is to reach 49% by the year 2020 [5].

According to the statistics published at the web page of District Heating in Sweden (Svensk Fjärrvärme), a total of 57.0 TWh was delivered as heat and 8.4 TWh as electricity from our district heating facilities in the year 2011. This corresponds to 6.0 and 0.9 MWh per capita, respectively. District heating accounts for about half of the domestic heating in Sweden. The contributions from various sources are shown in fig. 2.

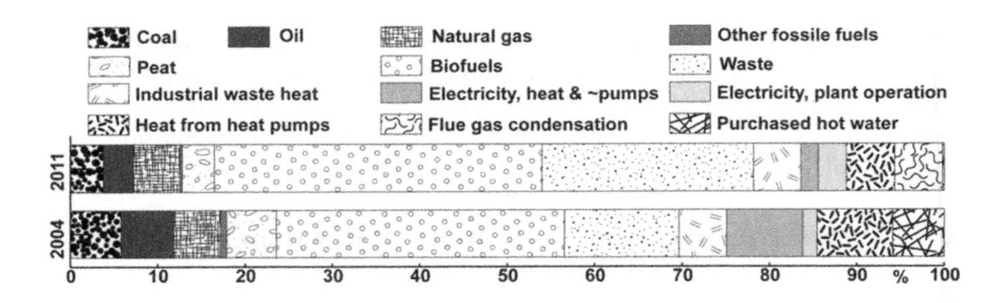

Figure 2: The contributions from the various sources for the heat and electricity supplied by the district heating plants in Sweden.

The efficiency in the heat production is 97%, not including the flue gas condensation. The efficiency is somewhat lower for electricity generation since other sinks, apart from the district heating grids, may sometimes be used for condensation of the steam e.g. in the summer when the need for heat is low. The high efficiency includes the incineration of waste.

Each year, the Swedish Waste Management (Avfall Sverige) publishes a report with statistics on the waste management [6]. Detailed data is provided on the emissions from the incineration facilities and it is found that they are well within the legal limits and the various permits. All incineration in Sweden constitutes energy recovery of the waste as defined in Annex II of the EC Waste Directive [7].

In the year 2011, a total of 2.2 million tons of domestic waste and about 3 million tons of other waste was incinerated, generating a total of 13.5 TWh of heat and 2 TWh of electricity.

Ashes from combustion and incineration exit at different points in the facilities and in different forms and this renders them somewhat difficult to categorize. It should be observed that the ash content of different fuels may vary considerably. It may be as low as 0.5% in stem wood and as high as 30% in domestic waste. The annual total generation of ash, including district heating, wood and the paper and pulp industry has been estimated to 1.48 million tons [8].

Avfall Sverige reports [6] that 880 thousand tons of bottom ash and 257 thousand tons of fly ash are generated from incineration of waste. Most of the bottom ash is used for covers of landfills while most of the fly ash is hazardous waste which is either deposited in landfills or sent abroad for stabilization. This is in concordance with the experience of the first author who has analyzed and classified ashes from more than two thirds of the incineration furnaces [2].

According to ref. [8], only 2% of all the ash generated is returned to the forest as a nutrient.

Landfilling in Sweden has decreased from 6.08 million tons in the year 1994 to only 1.52 in 2011 [6]. This implies that the need for ash for covers of landfills will decrease accordingly.

3 Legal aspects

3.1 Legislation

Legislation in Sweden comprises three levels: laws, ordinances and regulations. They are issued by parliament, government and regulatory authorities, respectively. Some issues are covered by EC regulations, and they are directly applicable in Sweden. EC directives have concomitant national legislation. In the following, the directive corresponding to the national legislation in question will be discussed for simplicity.

3.1.1 Some general considerations in the Swedish environmental code

The Swedish environmental legislation is regulated in a comprehensive manner by the Swedish environmental code [9]. Some "general rules of consideration" are presented in table 1. The following is of particular interest in the present context:

Table 1: Some general rules of consideration from chapter 2 in the Swedish environmental code [9].

Section 2	"*Persons who pursue an activity or take a measure, or intend to do so, must possess the knowledge that is necessary in view of the nature and scope of the activity or measure to protect human health and the environment against damage or detriment.*"
Section 3	"*Persons who pursue an activity or take a measure, or intend to do so, shall implement protective measures, comply with restrictions and take any other precautions that are necessary in order to prevent, hinder or combat damage or detriment to human health or the environment* ...".
Section 3	"... *the best possible technology shall be used* ...".
Section 5	"*Persons ... shall conserve raw materials and energy and reuse and recycle them wherever possible. Preference shall be given to renewable energy sources.*"

3.1.2 The Swedish environmental code and the waste framework directive

Some of the content of the EC waste framework directive [7] is implemented in chapter 15 of the Swedish environmental code [9], including the following.

Article 5 of the EC waste framework directive [7] states when a production residue is a by-product and not waste:

"*(a) further use of the substance or object is certain;*

(b) the substance or object can be used directly without any further processing other than normal industrial practice;

(c) the substance or object is produced as an integral part of a production process; and

(d) further use is lawful ..."

Article 6 of the EC waste framework directive [7] states when waste has become a product:

"*(a)* *the substance or object is commonly used for specific purposes;*
(b) *a market or demand exists for such a substance or object;*
(c) *the substance or object fulfils the technical requirements for the specific purposes and meets the existing legislation and standards applicable to products; and*
(d) *the use of the substance or object will not lead to overall adverse environmental or human health impacts.*"

3.1.3 Environmentally hazardous activities

The *ordinance (SFS 1998-899) concerning environmentally hazardous activities and the protection of public health* regulates when reporting and applications for permits are required. In general, this is required for combustion and incineration facilities with a capacity exceeding 0.5 MW. It should be noted that the lowest level, reporting to the municipal board, includes the option for the board to forward the matter to higher instances.

3.1.4 Combustion and incineration

Incineration of waste as well as mixtures of waste and fuel products is carried out under the incineration directive [10]. The corresponding Swedish legislation includes the ordinance SFS 2002:1060 together with the regulations NFS 2002:26 and NFS 2002:28. They contain limits on releases together with requirements on continuous as well as recurrent monitoring for fuel that is waste as well as products.

The legislation is intended to ensure that limits on emissions are not to be exceeded even in cases where there are considerable variations in the composition of the fuel. The protection is accomplished through the concerted effect of the legislation just mentioned together with the permit issued in concordance with SFS 1998-899 (see Section 3.1.3) which includes data on the flue gas cleaning system and a list of fuel types to be used.

3.1.5 Ongoing work at the EC Commission

It has not escaped the attention of the EC Commission that there is a difference in degree of detail in the legislations on combustion of fuels that are products and that on incineration of waste. In concordance, the Commission has commissioned its Joint Research Centre to investigate the issue. It is the intention that the results will form the basis for new regulations. The Centre has, in turn, contracted work to the Environment Agency Austria who have published an interim report [11].

The report finds that the emission limits under the waste directive are stricter than what may apply for non-waste, and thus some further specification for the fuel is warranted in the latter case.

The report concludes that "product and end-of-waste standards/specification already exist or will be developed ..." for the following: biogas, biodiesel and bio-ethanol. It also finds that end-of-waste criteria should be possible to obtain for e.g. wood and plastics.

The report finds that "In case such criteria for the end-of-waste status have not been set at community level, Member States may decide case-by-case whether certain waste has ceased to be waste, taking into account the applicable case law." Such an attempt is described in the subsequent section.

3.2 Court cases

Two court cases have been found which relate to the question of waste or product: case M585-08 on combustion of a filter medium, and M7546-08 on a residue from reprocessing of paper. Both were settled at our highest environmental court, Svea Court of Appeal, the Land and Environmental Court unit. Only the latter case will be described in the following.

The issue was dealt with consecutively at the County Administration Board at Västerbotten, the Land and Environmental Court at Umeå, and the Svea Court of Appeal. The courts were supported by expertise statements issued by the Swedish Environmental Protection Agency (EPA). The court decisions were essentially the same in all of the rulings.

Court documents in Sweden are (with few exceptions) open to the public, and the acts in question have been analyzed for the purpose of the present work. However, only those aspects of the case that are relevant for the purpose of the present paper are summarized in the following.

The company, SCA Packaging Ubbola AB, intended to burn the residue together with bark in its furnace on site. Formally, bark is regarded as waste (at least in the EPA statement), but it is exempted from the rules on incineration. The main issue was whether or not the residue, containing about 30% of plastics, should be considered as waste.

The lawyer representing the company focused on the rulings in the Court of Justice of the EC, and not as much on the technical issues as did the County Administration Board supported by EPA. The latter maintained that the plastic residue contains heavy metals as well as chlorine which is a known promoter of dioxin formation. No real attempts were made by the lawyer to prove that classification of the residue as a product, under the specifications by the company, and under the legislation of the environmental code (see Section 3.1.1) and some other legislation such as CLP and REACH, would render at least as good a protection of health and environment as would be the case under the legislation on incineration.

Conversely, the EPA pointed out that classification as a product would provide no legal obstacle for the supervisory authority to prescribe whatever precautionary measures that it might find appropriate.

The ruling was that the residue is waste, and the company had to install the monitoring equipment required under the legislation on incineration.

4 Metabolism of combustion and incineration

4.1 Introduction

Modern systems and processes for combustion and incineration of renewable fuels are very complex and conditions vary considerably with regard to e.g. fuel, furnace design, flue gas cleaning, ash management, type of production (heat or heat and electricity), and load characteristics (base or peak load).

In consequence, all these areas have been the subject of intensive research and development activities at the plants, in collective branch research programs and at universities. The domestic work published during the last ten years or more has been reviewed for the purpose of the present work. However, only very brief summaries of some important facts and findings are given in the following, in most cases without references. The authors hope to return to this issue in a future publication.

4.2 The metabolism of combustion and incineration processes

Basic and general descriptions of combustion and incineration processes can be found in [12, 13]. Three different furnace types are used: grate fired systems, fluidized bed systems and furnaces with powder burners. Ash is collected as fly ash and bottom ash except for powder burners that generate only fly ash.

Furnace systems together with operation parameters are chosen such that all combustible material is oxidized and that all burnable contaminants (substances that may be detrimental to health and the environment) become destroyed. This implies e.g. that temperatures be kept higher and over times longer for incineration of fuels containing certain very stable chlorinated compounds. Incomplete combustion can also lead to formation of poly-aromatic hydrocarbons (PAH).

In some cases, dioxins can form, and the following parameters promote such generation: long time at intermediate temperature, presence of chlorine, presence of transition metals, and incomplete combustion. Modern furnaces operating with conditions unfavorable with regard to dioxin generation produce only very low levels. Moreover, if dioxin levels are low, so are usually also those for PAH.

Chlorine associates to other elements somewhat differently at different temperatures and depending on the water content. It readily associates itself with potassium and sodium as well as with calcium, but also with e.g. zinc and lead thus rendering them much more volatile than would otherwise have been the case. The latter elements can therefore cause problems in terms of scale on the tubes as well as high leach rates in the ash, at least before they are hydrolyzed and incorporated into other phases. Chlorine that is not balanced by metallic elements as just described leaves the furnace in the form of hydrochloric acid vapor. Only limited levels of this compound may be released and therefore hydrochloric acid may have to be removed either by wet scrubbing or flue gas condensation (or both).

The metabolism of sulfur is somewhat more complex since sulfur has the valence six at low temperatures and that of four at high. Consequently, sulfur dioxide has a tendency to oxidize at lower temperatures, but the reaction is inefficient in the gas phase and is much more rapid on surfaces containing oxides of certain transition metals. This implies that chlorides deposited on the surfaces of tubes are converted to hydrochloric acid as incoming sulfur dioxide is oxidized and this acid may cause corrosion on the surfaces of the tubes. There are limits on emissions of sulfur dioxide and therefore it may have to be removed by semi-dry or wet scrubbing. Flue gas condensation is usually not sufficient.

It is not only the absolute levels of sulfur and chlorine that is of interest, but also their relative abundance. Chlorides usually have comparatively low melting points and do not dissolve much of oxides. They do, however mix with sulfates, which have somewhat higher melting temperatures. Mixtures of chlorides and sulfates with moderately high chlorine to sulfur ratios melt at even lower temperatures than the corresponding chlorides. Consequently, the chlorine to sulfur levels are often kept sufficiently low in order to avoid scale formation.

In practice, this may mean that a fuel rich in chloride, such as domestic waste, may be mixed with some other fuel rich in sulfur, e.g. sewage sludge.

Transition metals and heavy metals are distributed between bottom ash and fly ash in relation to their volatility under combustion conditions. Most of these elements are enriched in the fly ash, but to a very varying degree.

For the most part, transition and heavy elements are efficiently removed by electrostatic filters and/or mechanical filters (or other filters with similar capability). This may not be the case, however, for certain very volatile elements, especially mercury, but also cadmium. Mercury might therefore have to be absorbed by e.g. active carbon.

The following elements are of particular interest with regard to the health and environment related properties of the ash: arsenic, antimony, lead, chromium, molybdenum, nickel and zinc. In many cases, chromium, nickel and zinc may be or become self-stabilized in the ash by solid solution in iron-rich phases [2].

Different fuels have different contents of these elements. For instance, most of the arsenic in the ash originates from CCA-impregnated wood (CCA = Copper, Chromium and Arsenic) and can appear in recycled wood. Lead may be associated with old paint and plastics and may be present in domestic waste as well as recycled wood.

Fuels of interest include animal waste, domestic waste, paper, peat, plastics and rubber, pure vegetable fuels, recycled wood, residues from the pulp and paper industries, sewage sludge and tall oil.

It is a prerequisite for efficient and qualified ash utilization that contaminants be concentrated in a portion of the ash. This can be achieved by a combination of efficient sorting of the fuel together with careful selection of the fuels to be used at the various facilities.

5 Discussion and conclusions

A system for thermo-chemical energy from renewable sources should be characterized by the following:

1) protection of health and the environment;
2) sustainable with regard to:
   (a) the fuels that should be renewable,
   (b) the substances exiting the system that should as far as is possible and reasonable be reused and re-circulated in the anthroposphere;
3) efficient and cost effective.

Protection of health and the environment is an absolute prerequisite. Legislation must be followed to letter and should be followed also to intent.

Efficiency and cost effectiveness are necessary in order for an activity to be feasible and possible to be carried out.

As can be seen from section 4.2 above, most of the fuels can be regarded as sustainable. This includes domestic waste and industrial miscellaneous waste where most of the heat comes from wood, either directly or in the form of paper. However, domestic waste also contains plastics, and the level of fossil carbon has been estimated to 12.6% by weight [14]. However, actual measurements of the content of carbon-14 in the waste carried out by Renova in Gothenburg indicate a level of only around 2%.

It should be considered, when the sustainability of incineration is to be assessed, that the practice implies that the vast majority of potentially harmful substances in the waste are destroyed. Economically, the value of the destruction corresponds approximately to the cost of virgin fuel of the same calorific value.

This should be kept in mind also when the waste hierarchy in the EC waste directive (Article 4) [7] is being considered. In addition to the destruction of potentially harmful substances, incineration also reduces volume, and often converts the waste to a form suitable for recycling or disposal.

Actually, the sustainability of the metabolism of matter in incineration and combustion systems depends on a number of factors. It is not only the figure for total utilization that should be used, but also quality factors for the various uses. For instance, merely choosing ash for preparing the surface of the waste in a landfill before applying the seal of a cover is a comparatively unqualified use for which even quite contaminated ash may be feasible. Utilization in a cover or in a road requires low contents of contaminants together with specific desirable materials properties. Conversely, it is important to find applications in which also elevated levels of contaminants pose no threat to health and the environment.

The possibilities for utilization of ash depend on the chemical and physical properties as well as the levels of contaminants, but the bottleneck is usually the latter, especially the levels of elements like lead, arsenic, antimony, zinc, chromium, molybdenum, nickel. It is not only their levels that are of interest but also their chemical form, and this has been dealt with in a previous publication [2].

The contaminants are not evenly distributed between the various waste streams, and to some extent the latter can be separated into streams with different levels of contaminants before incineration. On the other hand, mixing different streams of waste is often favorable for the combustion process and may lead to less favorable concentrations of contaminants from an ash recycling perspective.

If waste could be sorted to near virgin fuel quality, then the corresponding ash would be easier to utilize in a qualified manner. Such a comparison with virgin material is not a formal requirement (although frequently put forward such as was done in the case M7546-08, see Section 3.2). Nonetheless, such an approach makes good sense since it lowers the threshold for use of fuel as a product. Some fuels do not have a corresponding virgin fuel, e.g. plastics, but it is feasible to develop corresponding standards also in such cases.

The option to classify fuel as a product as presented in Section 3.1.2 increases the degrees of freedom for the companies and thereby also the possibilities for achieving and improving sustainability.

On its internet page, the Svea Court of Appeal includes the court case M7546-08 in a listing of cases that are important precedents. It is obvious from Section 3.2 that this includes that health and environment has to be protected to the satisfaction of the Authorities and the Courts regardless of whether the fuel is waste or a product. Classification of fuels as products should thus not be attempted as any easy fix to avoid responsibility.

The ruling should not, however, be interpreted to imply that classification as a product in accordance with Section 3.1.2 above is not possible in Sweden. Instead it is for the owners and operators of the facilities to decide whether they chose to go ahead with such classification now or if they prefer to await the further developments at the EC Commission. This conclusion is in concordance with the following statement in [11]:

"*In the current absence of EU-wide end-of-waste criteria [for specific waste types], Member States may decide case by case whether certain waste has ceased to be waste taking into account the applicable case law.*"

For such case-by-case classification it is important to realize that the legislation on waste incineration rests on a comprehensive knowledge base and that adequate proof will have to be provided in the case of classification as a product, e.g. on variations together with quality control and assurance.

It is concluded that Sweden is champion, at least in Europe, in the utilization of renewable energy, and that this practice is also sustainable. However, the systems are complex and improvements are possible with regard to the utilization of substances exiting the system. This requires amongst other things further integration between the different areas of expertise: fuel, combustion and ash utilization, as well as between lawyers, engineers and scientists.

Acknowledgements

The present work is based on more than ten years of research reports (mostly in Swedish) financed by *District Heating in Sweden* (*Svensk Fjärrvärme*), [The Swedish] *Thermal Engineering Research Institute* (*Värmeforsk*), the *Swedish*

Waste Management (*Avfall Sverige*) and *Svenska Energiaskor AB* (which translates to: "Swedish Energy Ashes Inc."). Much of the present analyses have been made as a part of a recent commission from District Heating in Sweden.

References

[1] Baccini, P. & Brunner, P.H., *Metabolism of the anthroposphere: analysis, evaluation, design.* MIT Press, 2nd edition, 2012.

[2] Sjöblom, R., Classification of waste as hazardous or non-hazardous – the cases of ash and slag. *WIT Transactions on Ecology and The Environment*, Vol. 163, pp. 285–296, 2012.

[3] *Annual Energy Balance Sheets 2010 – 2011.* Statistics Sweden, Statistiska meddelanden, EN 20 SM 1206, korrigerad version, December 2012. Published on commission by the Swedish Energy Agency.

[4] Lindskog, S, Labor, B. & Sjöblom, R., Sustainability of nuclear energy with regard to decommissioning and waste management. *International Journal of Sustainable Development and Planning*, in print.

[5] Directive 2009/28/EC of the European Parliament and of the Council of 23 April 2009 on the promotion of the use of energy from renewable sources.

[6] *Swedish Waste Management 2012.* Available at www.avfallsverige.se.

[7] Directive 2008/98/EC of the European Parliament and of the Council of 19 November 2008 on waste and repealing certain Directives.

[8] *Ashes in Sweden 2010* (Swedish title: Askor i Sverige 2010). Published by Svenska Energiaskor AB and available at www.energiaskor.se.

[9] *The Swedish Environmental code* (translation of SFS 1998:808, not updated). Available at http://www.government.se/sb/d/2023/a/22847.

[10] Directive 2000/76/EC of the European Parliament and of the Council of 4 December 2000 on the incineration of waste.

[11] Stoiber, H., project manager (many authors), *Study on the suitability of the different waste-derived fuels for end-of-waste status in accordance with article 6 of the waste framework directive*, second interim report. Draft for consultation. Environment Agency Austria, Vienna, August 2011.

[12] Raask, E., *Mineral impurities in coal combustion behavior, problems, and remedial measures.* Hemisphere Pub. Corp. in Washington, 1985.

[13] Chandler, A.J., Eighmy, T.T., Hartlén, J., Hjelmar, O., Kosson, D.S., Sawell, S.E., Van der Sloot, H.A. & Vehlow, J., *Municipal Solid Waste Incinerator Residues*. Elsevier Science B.V. 1997.

[14] Sundqvist, J-O., *Evaluation of Swedish waste policy in a system perspective.* (Swedish title: Utvärdering av svensk avfallspolitik i ett systemperspektiv). Swedish Waste Management (Avfall Sverige) Report 2007:10.

An environmental balance study for the contribution of a biomass plant in a small town in Piedmont, Northern Italy

D. Panepinto & G. Genon
DIATI Department, Politecnico di Torino, Italy

Abstract

In consideration of local critical aspects in opposition to overall environmental benefits (decrease of greenhouse gas generation), the aim of this work is to verify the local acceptability of a biomass plant, from the point of view of air quality in the territory of interest. The plant, to be realized in a small town located in Piedmont, northern Italy, will be constructed to produce electricity and heat. In order to verify the aspect of compatibility, an evaluation is performed of the emissive flow modification that, on the hypothesis of the biomass plant activation, should be introduced in the municipal area. The evaluation has been conducted using mass and energy balances and a pollutant dispersion model as tools.
Keywords: biomass plant, district heating, environmental balance, environmental impact, energy recovery.

1 Introduction

Climate change is the major planetary threat of the XXI century, not only for the natural ecosystems but for some national economies too. In order to reduce immediately the emissions of greenhouse gas, it is strongly proposed to use renewable and clean sources instead of fossil fuels. In this research, one of the most important solutions consists in a modern use of the biomass [1, 2].

Fossil fuel such as coal, oil and natural gas are generating a very large proportion of the electricity produced annually around the globe. The combustion of these fuels gives rise to carbon dioxide (CO_2), which is a "greenhouse" gas discharged into the atmosphere. There is a growing interest in eliminating, as much as possible the CO_2 emissions from fossil sources.

In comparison, the carbon dioxide generated in the combustion of biofuels is not considered to give any net contribution to the CO_2 content of the atmosphere, since it is absorbed by photosynthesis when new biomass is growing [3]. Biomass is widely considered to be a major potential fuel and renewable resource for the future [4–9]. In this sense energy from biomass based on short rotation forestry and also from the exploitation of other energy crops can contribute significantly towards the objectives of the Kyoto Agreement in reducing the greenhouse gases emissions and consequently the problems related to climate change [10, 11].

In this work, the realization, in a little community in Piedmont (north Italy), of a biomass plant for energy generation is studied, with the finality to operate in conditions of cogeneration by producing electric energy to be fed into the electricity network and thermal energy, that is destined to satisfy the local requirements with a district heating network. The catchment area of interest is the little municipal area where the plant will be located.

In order to verify the environmental compatibility of the biomass plant, an evaluation of the emissive fluxes modification is performed at local level, using environmental balance and the implementation of pollutant dispersion model as a tool.

The new emissive flux has been considered; this can be predicted in direct relation to the biomass plant activation as a result of the thermal power plant, the type of fuel used and the system that will be employed for the environmental impact containment. On the other hand, the avoided emission flux has been evaluated, resulting from the turn off of the domestic boilers whose output could be substituted from the produced thermal energy.

The calculations that have been performed are based on the knowledge of the emission factors for different plant design solutions and this is established by considering also the thermal power of the used systems; mass and energy balances and the implementation of pollutant dispersion model which were used as tools in this study [12].

2 Biomass plant

In tables 1 and 2 the main features (concerning technical and energetic aspects) of the studied biomass plant are reported.

Table 1: Main features of the studied plant.

Fuel		Biomass – wood pellets
Technology		Combustion on grate system
Energy recovery		boiler
Availability		7,800 h/y
Treatment emissions	Dust	ESP (Electrostatic Precipitation)
	NO_x	SNCR (Selective Catalytic Reduction)
	SO_x	Injection of CaOH

In table 2, the main energetic plant features are reported.

Table 2: Summary of the biomass plant energetic data.

Gross thermal power	14.6 MW
Available thermal power	13 MW
Maximum thermal power	7.15 MW
Maximum required thermal power*	6.8 MW
*this data represent the maximum required thermal power (by considering all the public and private users present in the municipal area)	

By analyzing the data reported in table 2, it can be observed that the maximum thermal power produced from the plant (7.15 MW) cannot be fully utilized even if all the users of the whole town (6.8 MW) were connected.

The following table reports the pollutant concentration that, from the technological operating scheme, can be estimated as output of the plant.

Table 3: Pollutant concentration as output of the plant.

Flow [m³/h]	**Emission time [h/d]**	**Pollutant substance**	**Concentration [mg/m³]**
33.000	24 - continue	Dust	10
		NO_x	180
		SO_x	200
		CO	200

3 Catchment area and connectable volumes

The catchment area will be composed of the municipal area where the biomass plant will be located. Then, all the data that has been found have been summarized in a comprehensive table which is depicted below.

Table 4: Town catchment area.

Population	770
Number of resident families	394
Number of total inhabitations	983
Surface of resident inhabitation [m²]	43,965
Volume of resident inhabitations [m³]	131,965
Total inhabitations volume (resident + not resident) [m³]	390,000

From this, on the basis of data found and their aggregation, the volumes for connection to the future district heating network can be defined. These volumes are shown in table 5.

Table 5: Volume connectable to the district heating network.

Resident connectable volume [m^3]	81,818
Total connectable volume (resident + not resident) [m^3]	300,000

In the performance of the environmental compatibility analysis, four different situations were examined referring to four different scenarios of connection to the district heating system:

- hypothesis 1: the entire volume of the analyzed town (public and private utilities) will be connected to the district heating network (no distinction between residents and non residents, and no consideration of the aspect of volumes effective capacity to be connected or not);
- hypothesis 2: only the volumes of the public and private buildings that can be really connected to the district heating network with an acceptable cost will be considered without distinction between resident and non-residents;
- hypothesis 3: the total volume of residents will be connected to the district heating network (no distinction between effectively connectable and not connectable);
- hypothesis 4: the total volume of residents effectively connectable will be connect to the district heating network.

On the basis of the data reported in tables 4 and 5, the thermal energy really transferred was calculated for each introduced hypothesis.

Table 6: Really distributed thermal power.

	Hypothesis 1	Hypothesis 2	Hypothesis 3	Hypothesis 1
Heat specific requirement [kWh/m^3 y]	44	44	44	44
Period of heat requirement [h/y]	2,500	2,500	2,500	2,500
Required thermal power [kW_{th}]	**6,864**	**5,280**	**2,323**	**1,440**

The boilers that are installed in the homes located in the examined town are fed in part (the majority) by methane, and in part (a small part) by wood. On the basis of the data that were supplied by the local authority, it has been possible to establish, for each of the four analyzed hypotheses, the fuel composition (with consideration of the percentages of methane and wood) and, consequently, the thermal power that today is covered by methane and wood. The results of this task are presented in table 7.

Table 7: Definition of the boiler input and of the covered thermal power.

Hypothesis	Boiler feed	Satisfied thermal power [%]	Satisfied thermal power [kW_{th}]
Hypothesis 1	Methane	95	6,521
	Wood	5	343
Hypothesis 2	Methane	95	5,016
	Wood	5	264
Hypothesis 3	Methane	99	2,300
	Wood	1	23
Hypothesis 4	Methane	99	1,426
	Wood	1	14

4 Energetic and environmental balance

In order to evaluate the local environmental benefits, it is necessary to compare the air quality around the assumed CHP (Combined Heat and Power) location before and after installation of the DH (District Heating) system. That is a consequence of the modified emissions scenario. Therefore it is necessary to estimate the contribution of the existing boilers to the air emissions.

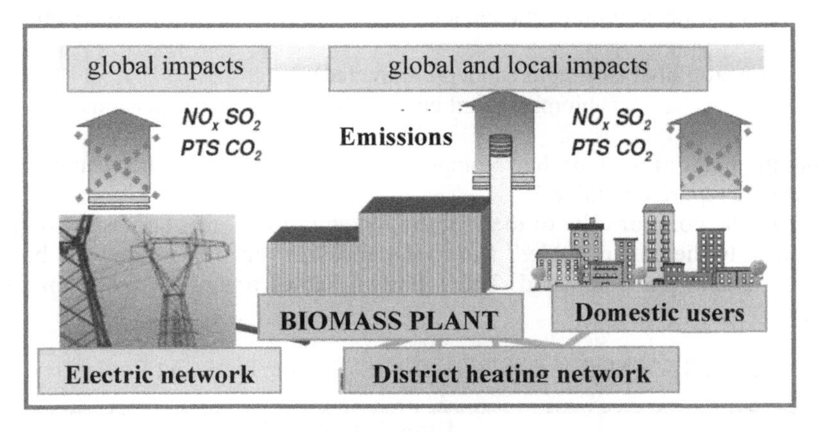

Figure 1: Environmental balance.

The electric energy distributed by the network substitutes part of the centralized electric production, and so, the relative environmental impacts expressed in terms of primary energy consumption and atmospheric emissions are avoided. The quantification of this impact derives from the considered comparison terms. At the same time the thermal energy supplied by a DH system allows the substitution of the operation of existing boilers and the relative impacts, such as primary energy consumption and atmospheric emissions. In this case the avoided impacts in an unequivocal way correspond to those of the impacts effectively substituted. In the draft of an environmental balance the two components of the avoided impacts represent compensation to the environmental load that is introduced by the DH system.

Besides the energetic and environmental balance and in order to evaluate the momentum of the produced impact of the plant, and for this estimation also by taking into account the aspect of its localization, it is necessary to consider the results of the dispersions models. With this approach it is possible to calculate the real air quality modifications; the concentrations (annual mean values and maximum hourly values) that are introduced by the future plant and the ones (due to elimination of existing domestic boilers) that can be avoided must be compared on the basis of concentration maps [13].

In general and with a first qualitative approach it is possible to consider that a single plant against many individual boilers can be advantageous in many respects, among which the minor pollution, specific production capacity and the major energetic efficiency. A great plant should have higher thermal efficiency and better smoke control in comparison with the performances of small plants. Also the aspect of higher emissions point of the chimneys, and of higher atmospheric emissions flow-rates and consequent higher dispersion capacities must be taken into account, as an aspect of advantage for concentrated solutions.

4.1 Environmental balance: initial considerations

The basis for the environmental balance is provided by the following equation:

$$\text{Local/global emissions (added/eliminated)} = \text{biomass plant emissions} - \text{substituted emissions} \quad (1)$$

From the point of view of the "biomass plant emissions", reference can be made to the data reported in table 3.

From the point of view of the "substituted emissions", it is necessary to make reference to the data of table 7 and to the emission factors for domestic boilers fed by natural gas and wood. The following tables report the used emissions factors.

Table 8: Methane emission factor. (Source: Piedmont Region.)

Fuel	Emission factor [mg/MJ]				
	Dust	NO_x	SO_x	CO	COV
Methane	1.98	50	0.51	25.25	4.64

Table 9: Wood emission factor. (Source: Piedmont Region.)

Fuel	Emission factor [g pollutant/kg fuel]				
	Dust	NO_x	SO_x	CO	COV
Wood	4.91	2.3	0.34	50	8.24

Using eqn (1), an environmental balance is obtained for each of the considered scenarios. The pollutant parameters that were considered in the estimation were:

- dust;
- nitrogen oxides (NO_x). For this pollutant parameter, two different situations were considered:
  - NO_x emission treated with $DeNO_x$ SNCR system as considered by the plant designers (data in table 3),
  - NO_x emission treated with $DeNO_x$ SCR system; in this case, in comparison with an higher investment cost there is an improvement in the performance as regards the removal of pollutant (in fact with an SNCR system the estimated pollutant concentration in output corresponds to 180 mg/m^3, as reported in table 3, while with an SCR system it is possible to arrive to a pollutant concentration in output of 108 mg/m^3);
- sulphur oxide (SO_x);
- carbon monoxide (CO);
- Volatile Organic Carbon (VOC).

4.2 Environmental balance on annual scale

The initial point was the evaluation of the hypothesis 1. According to this hypothesis, it is foreseen that all the town volume will be connected to the district heating network. The results of the calculations are reported in table 10.

Table 10: Environmental balance, hypothesis 1.

	Biomass plant (+)	Methane boilers (-)	Wood boilers (-)	Environmental balance
Dust [t/y]	2.5	0.12	1.21	+ 1.17
NOx (SNCR) [t/y]	46.3	2.93	0.57	+ 42.8
NOx (SCR) [t/y]	27.8	2.93	0.57	+ 24.3
SOx [t/y]	51.5	0.03	0.08	+ 51.39
CO [t/y]	51.5	1.48	12.35	+ 37.67
VOC [t/y]	5.15	0.27	2.035	+ 2.85

By analyzing the results of table 10, it is seen that environmental impact of the biomass plant will be higher in comparison with that of the substituted domestic boilers. It can be observed that this impact is conspicuous in particular for the parameters NO_x (both considering a SNCR system or a SCR system), SO_x and CO.

As for hypothesis 2, in this case it is considered that only the effectively connectable town volume will be connected to the district heating network. Table 11 reports the results of this hypothesis.

By analyzing the results, it can be observed that also in this case, the pollutant load introduced by the biomass plant is significantly higher in comparison with

Table 11: Environmental balance, hypothesis 2.

	Biomass plant (+)	Methane boiler (-)	Wood boiler (-)	Environmental balance
Dust [t/y]	2.5	0.09	0.94	+ 1.47
NOx (SNCR) [t/y]	46.3	2.24	0.44	+ 43.62
NOx (SCR) [t/y]	27.8	2.24	0.44	+ 25.12
SOx [t/y]	51.5	0.022	0.064	+ 51.4
CO [t/y]	51.5	1.14	9.5	+ 40.86
VOC [t/y]	5.15	0.21	1.57	+3.37

the avoided one arising from the substitution of the domestic boilers that are effectively connectable to the district heating network. As with the previous hypothesis, the parameters that mainly suffer from the biomass plant introduction are the NO_x, the SO_x and the CO; also the parameters dust and VOC undergo a worsening at balance level, but to a lesser measure.

Hypothesis 3 is based on the assumption that only the volume corresponding to residents will be connected to the district heating network. In table 12, the result referred to this hypothesis is reported; it can be seen that the trend is similar to the trend in the previous situations.

Table 12: Environmental balance, hypothesis 3.

	Biomass plant (+)	Methane boiler (-)	Wood boiler (-)	Environmental balance
Dust [t/y]	2.5	0.04	0.08	+ 2.38
NOx (SNCR) [t/y]	46.3	1.04	0.04	+ 45.22
NOx (SCR) [t/y]	27.8	1.04	0.04	+ 26.72
SOx [t/y]	51.5	0.01	0.006	+ 51.48
CO [t/y]	51.5	0.52	0.83	+ 50.15
VOC [t/y]	5.15	0.1	0.14	+ 4.91

Hypothesis 4 shows that only the volume for residents effectively connectable will be connected to the district heating network. In table 13, the results relative to this hypothesis is reported.

Table 13: Environmental balance, hypothesis 4.

	Biomass plant (+)	Methane boiler (-)	Wood boiler (-)	Environmental balance
Dust [t/y]	2.5	0.025	0.05	+ 2.43
NOx (SNCR) [t/y]	46.3	0.64	0.023	+ 45.64
NOx (SCR) [t/y]	27.8	0.64	0.023	+ 27.14
SOx [t/y]	51.5	0.006	0.0034	+ 51.49
CO [t/y]	51.5	0.32	0.504	+ 50.68
VOC [t/y]	5.15	0.06	0.083	+ 5

As expected, in this case the load introduced by the plant is much higher than the subtracted ones arising from shutdown of domestic boilers connectable to the district heating network.

4.3 Percentage of pollutant increase with consideration of background value

In the previous sections, the evaluated pollution loads were those introduced by the future biomass plant and those which may be eliminated by the substitution of a number of public and private boilers.

In these calculations, the background load deriving from other existing sources operating in the area, as in particular agricultural activities, animal farms, production operations, transport, was not considered.

Table 14 reports the contribution that can be estimated for these emission sources for the examined town, with consideration of the principal categories.

Table 14: Emissive contribution and subdivision for emission sources.

	Emissive contribution expressed in t/y				
	NO_2	PM_{10}	NH_3	CO	SO_2
Agriculture	2.16	0.28	1.03	–	–
Zootechnic	–	–	8.66	–	–
Productive activity	0.10	0.02	0	–	–
Transport	2.88	0.87	0.05	–	–
Urbanization	2.84	1.32	0	–	–
TOTAL (t/y)	**7.98**	**2.49**	**9.74**	**54.49**	**0.44**

The percentage increase, for each evaluated hypothesis, has been calculated using the following equation:

$$\% \text{ of increase} = ((\text{introduced load} - \text{substituted load})/\text{background value})*100 \quad (2)$$

with

$$\text{introduced load (t/y)} = \text{biomass plant emission} + \text{background value}$$

The obtained results are reported in the following figures.

If the results reported in fig. 2 are analyzed for all the scenarios and taking into account each pollutant parameter, an important percentage increase deriving from the activation of the new biomass plant can be noted. Among the five pollutants, the parameters that present a smaller percentage increase are the dust and the carbon monoxide. On the contrary the pollutant parameter SO_x presents the highest increase percentage among the five pollutant parameters examined. From the point of view of the pollutant parameter NO_x, there is a difference in the increase percentage in function of the adopted pollutant removal system

(SNCR or SCR). If a SNCR removal system is used, the increase percentage is on the order of 400 to 500%, while in case of using a SCR, the increase is on the order of only 200 to 300%.

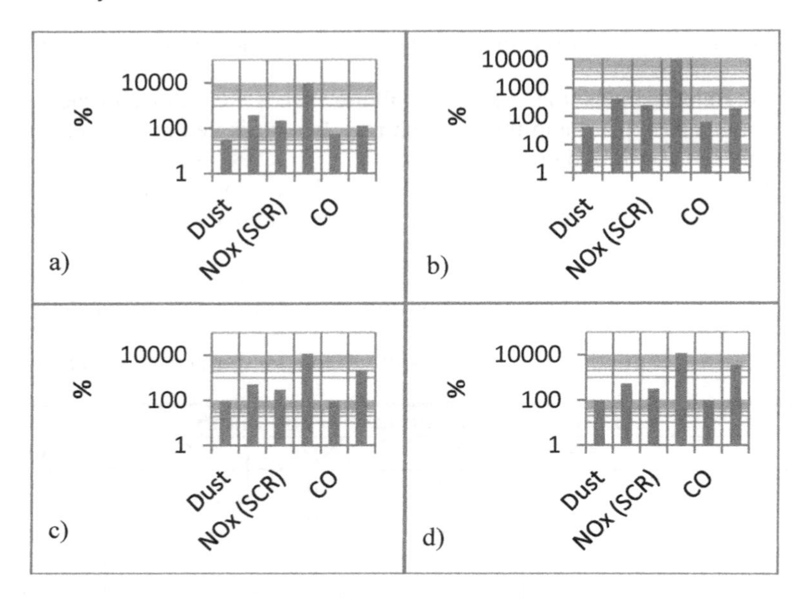

Figure 2: Percentage of pollutant increase with background value: a) hypothesis 1; b) hypothesis 2; c) hypothesis 3; d) hypothesis 4.

5 Implementation of pollutant dispersion models

In order to evaluate the severity of the environmental impact produced by the plant, it is necessary to consider the results of the dispersion models. With this approach it is possible to calculate the real air-quality modifications: the concentrations (annual mean values and maximum hourly values) that will be created by the future plant and the elimination of concentrations corresponding to the sources that will be avoided (from the elimination of existing domestic boilers) are compared. This comparison was performed by constructing concentration maps [13].

The maps were constructed from the results of the simulation of the atmospheric dispersion of pollutants emitted from all relevant sources, using the Aermod model. It is a Gaussian model, which uses the Gauss function of errors as an analytical solution of the equation of transport in the atmosphere [14].

5.1 Effects on air quality

In the following figure, the results of the pollutant dispersion analysis can be seen. In this case, only the results corresponding to the hypothesis 1 are indicated (because the other results obtained are quite similar).

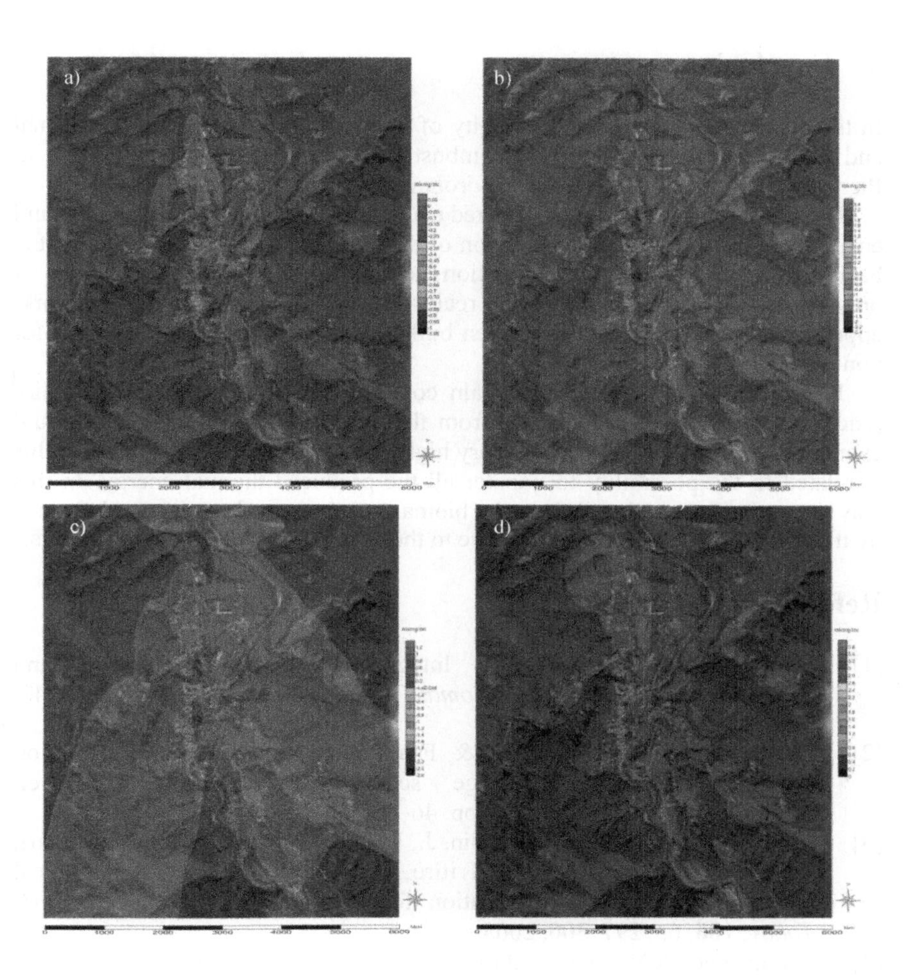

Figure 3: Pollutant dispersion results, parameters: a) dust; b) NOx (with SNCR); c) NOx (with SCR); d) SOx.

By analyzing the results reported in fig. 3, it is possible to see, from the point of view of the dust parameter, a general improvement of air quality. Also from the point of view of the parameter NOx, an improvement of air quality is observed; this improvement is greater in the case of adopting a Selective Catalytic Reduction (SCR) system for the reduction of the nitrogen oxide. From the point of view of the parameter SOx, instead, there is a general worsening of the air quality consequent to the activation of the plant.

6 Conclusions

In the presented work, the acceptability of a plant for the generation of thermal and electric energy with biomass combustion, to be realized in a little city in Piedmont, was evaluated from the environmental point of view.

This acceptability has been evaluated using the tools of the environmental and energy balance and the implementation of the pollutant dispersion model, on the basis of some hypothesis of connection (total volume of existing buildings or only volume of the buildings that really can be connected to the network, appreciation of the difference between buildings for residents and buildings for non residents).

From the obtained results the main conclusion is that an effective facility (and so an effective acceptability), from the environment point of view, could eventually be obtained only with a very high value of energy utilization, and this is related to the possibility to transfer all the produced thermal energy. In this way the emissions produced from the biomass plant are at least in part balanced by the values of avoided emissions, due to the turn off of some domestic boilers.

References

[1] Boman, U.R. & Turnbull, J.H., Integrated biomass energy systems and emissions of carbon dioxide. *Biomass and Bioenergy*, **13(6)**, pp. 333–343, 1997.

[2] Dornburg, V., Van Dam, J. & Faaij, A., Estimating GHG emission mitigation supply curves of large – scale biomass use an a country level. *Biomass and Bioenergy*, **31(1)**, pp. 46–65, 2007.

[3] Albertazzi, S., Basile, F., Brandin, J., Einvall, J., Hulteberg, C., Fornasari, G., Rosetti, V., Sanati, M., Trifirò, F. & Vaccari, A., The technical feasibility of biomass gasification for hydrogen production. *Catalysis Today*, **106**, pp. 297–300, 2005.

[4] Bridgwater, A.V., The technical and economic feasibility of biomass gasification for power generation. *Fuel*, **74**, pp. 631–653, 2005.

[5] Caputo, A.C., Palumbo, M., Pelagagge, P.M. & Scacchia, F., Economics of biomass energy utilization in combustion and gasification plants: effects of logistic variables. *Biomass and Bioenergy*, **28(1)**, pp. 35–51, 2005.

[6] Hanaoka, T., Inove, S., Uno, S., Ogi, T. & Minowa, T., Effect of woody biomass components on air – steam gasification. *Biomass and Bioenergy*, **28(1)**, pp. 69–76, 2005.

[7] Hohenstein, W.G. & Wright, L.L., Biomass energy production in the United States: an overview. *Biomass and Bioenergy*, **6(3)**, pp. 161–173, 1994.

[8] Hustad, J., Skreiberg, Ø. & Sonju, O., Biomass combustion research and utilization in IEA countries. *Biomass and Bioenergy*, **9(1-5)**, pp. 235–255, 1995.

[9] Van Den Broek, R., Faa Ij, A. & Van Wick, A., Biomass combustion for power generation. *Biomass and Bioenergy*, **11(4)**, pp. 271–281, 1996.

[10] Maniatis K., *Progress in Biomass Gasification: An Overview*, Directorate General for Energy & Transport, European Commission, 2002.

[11] IEA Bioenergy, The role of bioenergy in greenhouse gas mitigation, Position paper, IES Bioenergy, New Zealand, 1998.

[12] Panepinto, D. & Genon, G., Environmental balance study for the construction of a biomass plant in a small town in Piedmont (Northern Italy). *WIT Transactions on Ecology and the Environment*, **143**, pp. 279–290, 2011.

[13] Genon, G., Torchio, F.M., Poggio, A. & Poggio, M., Energy and environmental assessment of small district heating systems: global and local effects in two case-studies. *Energy Conversion and Management*, **50**, pp. 522–529, 2009.

[14] Panepinto, D. & Genon, G., Biomass thermal treatment: energy recovery, environmental compatibility and determination of external costs. *Waste and Biomass Valorization*, **3**, pp. 197–206, 2012.

Author index

CPSIA information can be obtained at www.ICGtesting.com
Printed in the USA
BVOW11*0840190515

399522BV00003B/4/P